Enantioselection in Asymmetric Catalysis

Enantioselection in Asymmetric Catalysis

Ilya D. Gridnev

Pavel A. Dub

CRC Press
Taylor & Francis Group
Boca Raton London New York

CRC Press is an imprint of the
Taylor & Francis Group, an **informa** business

CRC Press
Taylor & Francis Group
6000 Broken Sound Parkway NW, Suite 300
Boca Raton, FL 33487-2742

First issued in paperback 2019

© 2017 by Taylor & Francis Group, LLC
CRC Press is an imprint of Taylor & Francis Group, an Informa business

No claim to original U.S. Government works

ISBN-13: 978-1-4987-2654-2 (hbk)
ISBN-13: 978-0-367-87325-7 (pbk)

Library of Congress Cataloging-in-Publication Data

Names: Gridnev, Ilya D., 1963- | Dub, Pavel A.
Title: Enantioselection in asymmetric catalysis / Ilya D. Gridnev and Pavel A. Dub.
Description: Boca Raton : CRC Press, 2017. | Includes bibliographical references and index.
Identifiers: LCCN 2016032429 | ISBN 9781498726542 (hardcover : alk. paper)
Subjects: LCSH: Enantioselective catalysis. | Catalysis. | Asymmetric synthesis.
Classification: LCC QD505 .G7545 2017 | DDC 541/.395--dc23
LC record available at https://lccn.loc.gov/2016032429

Visit the Taylor & Francis Web site at
http://www.taylorandfrancis.com

and the CRC Press Web site at
http://www.crcpress.com

Contents

Introduction

The idea of writing this book came when the authors were finalizing their long-term studies of the mechanisms of asymmetric Rh-catalyzed hydrogenation of activated olefins (IDG) and Ru-catalyzed hydrogenation and transfer hydrogenation of aromatic ketones (PAD). Both fields have attracted a considerable interest from researchers due to the utmost effectiveness and extremely high levels of enantioselectivity that were achieved in certain reactions of this type. This supports the attempts to uncover the real mechanism of the generation of chirality, because the observation of optical yields of over 99% ee suggests it would be easier to locate a real catalytic pathway among the numerous possibilities.

Nevertheless, during their experimental and computational studies, the authors became convinced that, even in the case of extremely high optical yields, it is not easy to establish reliably the actual pathways taken in the multistep catalytic cycles and elucidate the factors important for the effective generation of chirality.

There are several reasons for these complications, but all of them come from the same origin: the multistep character of the catalytic cycles. It means, for example, that several stages of a chiral catalytic cycle can be enantioselective. Therefore, we can compute or even observe experimentally a high enantioselectivity for a certain step of the catalytic cycle but still cannot be completely confident if it is actually this step that is responsible for the overall chirality of the product.

Practically it means that the whole network of possible catalytic pathways that may be converging and/or may contain bifurcation points must be computed before a reliable decision on the nature of the enantioselective step can be made. This, in turn, requires accurate knowledge of all possible experimental data on the relative stabilities of the intermediates, structure of the resting state, kinetics, etc., to calibrate and control the computational results.

In other words, the elucidation of the mechanism of enantioselection in an asymmetric catalytic reaction is not an easy and straightforward task; it requires extensive experimental and computational studies. The first two reactions discussed in this book are evidently exceptional with respect to the amount of research that has been used to elucidate the mechanism of these reactions. One can confidently claim no other catalytic asymmetric reaction can be compared in this respect to Rh-catalyzed asymmetric hydrogenation of activated olefins and Ru-catalyzed hydrogenation and transfer hydrogenation of ketones. However, despite the research carried out in this field over several decades, some problems still remain unresolved.

Unfortunately this means the research of the mechanism of enantioselection of many other reactions presented in this book can hardly be considered complete, and this problem has recently been discussed using the Morita–Baylis–Hillman reaction as an example.[1]

Taking into account all these complications, this book does not attempt to give a comprehensive compilation of the current research in this field. Rather, representative asymmetric catalytic reactions catalyzed by chiral complexes of transition metals or organocatalytic reactions not highlighted in the recent reviews[2,3] are discussed.

Although the authors tried to make the illustrations of the important transition states as clear as possible, this is not always possible for complex 3D structures. Hence, for the convenience of the readers, this book is accompanied by a CD containing the key structures in the ".mol" format that is readily readable by most commercial and freeware visualization software.

References

1. Plata, E. R.; Singleton, D. A. A Case Study of the Mechanism of Alcohol-Mediated Morita Baylis–Hillman Reactions. The Importance of Experimental Observations. *J. Am. Chem. Soc.* 2015, *137*, 3811–3826.
2. Sperger, T.; Sanhuesa, I. A.; Kalvet, I.; Schoenebeck, F. Computational Studies of Synthetically Relevant Homogeneous Organometallic Catalysis Involving Ni, Pd, Ir, and Rh: An Overview of Commonly Employed DFT Methods and Mechanistic Insights. *Chem. Rev.* 2015, *115*, 0532–9586.
3. Cheong, P. H.-Y.; Lagault, C. Y.; Um, J. M.; Çelebi-Ölçüm, N.; Houk, K. N. Quantum Mechanical Investigations of Organocatalysis: Mechanisms, Reactivities, and Selectivities. *Chem. Rev.* 2011, *111*, 5042–5137.

Authors

Ilya D. Gridnev earned a PhD in 1989 at Moscow University. From 1990 to 1998 he worked in Moscow at the Institute of Organic Chemistry and Institute of Organoelement Compounds at the Russian Academy of Science. After two postdoctoral fellowships, one in Japan (Japanese Society for the Promotion of Science Fellowship, Hokkaido University, Sapporo, Hokkaido) and the other in Germany (A. v. Humboldt Fellowship, Göttingen University, Göttingen), he earned a DrSc ("Habilitation") at the A. V. Nesmeyanov Institute of Organoelement Compounds, Moscow, Russia. From 1998 to 2007 he was a research associate at the University of Rennes (France), Chiba University (Japan), and Oxford University (United Kingdom). From 2003 to 2007 he was an associate professor at Tohoku University (Sendai, Japan) and from 2007 to 2012 at the Tokyo Institute of Technology. Since 2012 he has been an associate professor at Tohoku University, Sendai, Japan. His research interests include asymmetric catalysis, computational chemistry, organometallic chemistry, reaction mechanisms, and NMR spectroscopy.

 **Pavel A. Dub** earned an MS at the Higher Chemical College of the Russian Academy of Sciences in 2007 and two PhDs at the Russian Academy of Sciences and the Université de Toulouse in 2009 and 2010, respectively, working at the A. N. Nesmeyanov Institute of Organoelement Compounds (Professor Elena S. Shubina) and the Laboratoire de Chimie de Coordination, UPR CNRS 8241 (Professor Rinaldo Poli). He spent two and a half years (2010–2013) as a JSPS postdoctoral fellow in Professor Takao Ikariya's group at the Tokyo Institute of Technology. He is a technical staff member at the Los Alamos National Laboratory, Los Alamos, New Mexico, where he was Director's and then J. Robert Oppenheimer Distinguished Postdoctoral Fellow with Dr. John C. Gordon (2013–2016). His current research interests include organometallic chemistry, computational quantum chemistry, and practical molecular catalysis.

chapter one

Transition metal-catalyzed hydrogenation

1.1   Rh-catalyzed asymmetric hydrogenation of activated olefins

1.1.1   Overview

The discoveries of Osborn et al.[1,2] demonstrated in 1965 high effectiveness of soluble Rh complexes in catalytic hydrogenation of simple olefins. Only after 3 years, the first asymmetric variants of this reaction using chiral monophosphines as ligands were reported.[3,4] Then the value of chiral diphosphines that form chelating complexes with Rh has been recognized, and for next 20 years most of synthetic and mechanistic studies were carried out using C_2-symmetric Rh diphosphine complexes. Although a renaissance of the chiral monophosphine ligands has taken place in early 2000,[5–8] the diphosphine Rh complexes keep their importance until now. As a result, a tremendous number of chiral mono- and diphosphines have been tried as ligands in Rh-catalyzed asymmetric hydrogenation. Hence, mostly the mechanism of the asymmetric hydrogenation catalyzed by Rh–diphosphine complexes will be discussed here, and the catalysis by Rh–monophosphine complexes will be shortly reviewed afterward.

1.1.2   Asymmetric hydrogenation catalyzed by Rh–diphosphine ligands

1.1.2.1   Resting state of the catalytic cycle: Formation of the catalyst–substrate complexes

Unlike the Wilkinson catalyst that irreversibly yields a dihydride complex when treated with dihydrogen, the Rh–diphosphine complexes after hydrogenating off diene ligand from the catalytic precursor **1** give solvate complexes **2** (Scheme 1.1).[9]

The solvate complexes **2** are real catalysts that are recovered after each catalytic cycle. Reaction of solvate complexes with a substrate in most cases leads to the formation of catalyst–substrate complexes.[10] Although the binding is very fast and essentially barrierless,[11] it has been shown that initially

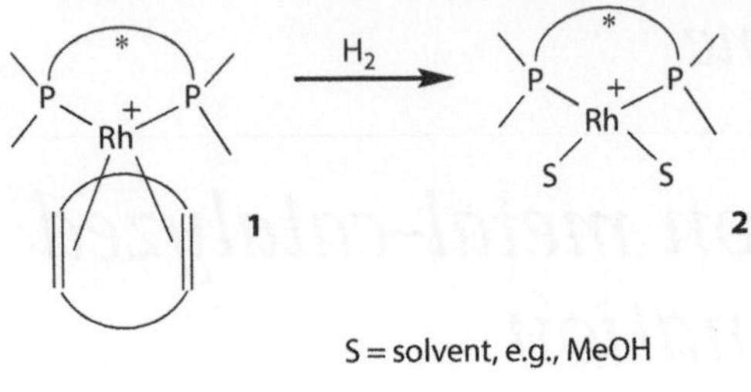

S = solvent, e.g., MeOH

Scheme 1.1 Formation of a catalyst **2** from a catalytic precursor **1**.

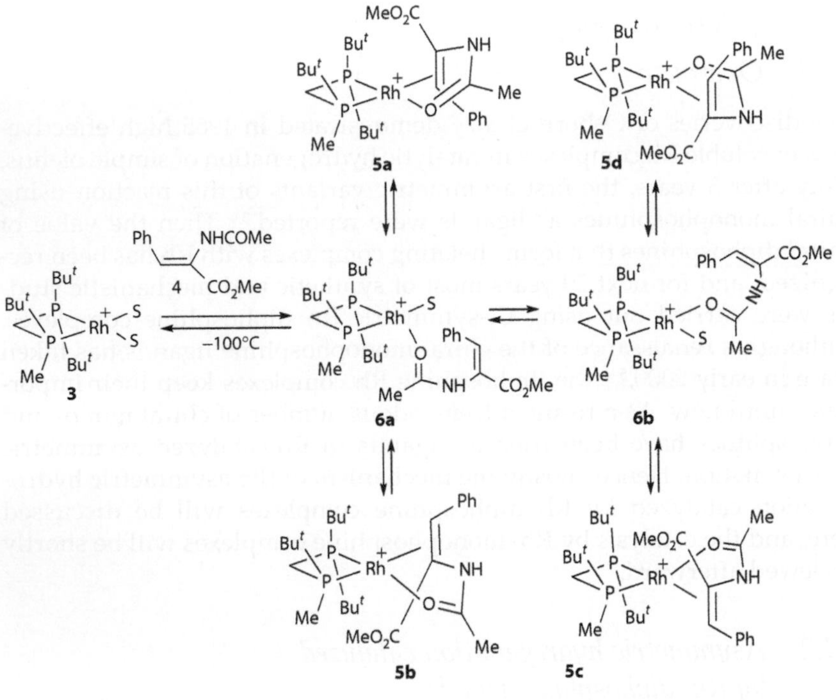

Scheme 1.2 Low-temperature reaction of catalyst **3** and substrate **4**. (Gridnev, I. D. and Imamoto, T., *Chem. Commun.*, 7447–7464, 2009. Reproduced by permission of The Royal Society of Chemistry.)

the amidocarbonyl group of the substrate forms a nonchelating complex (Scheme 1.2).[12]

When substrate **4** is added to catalyst **3** at −100°C, relatively unstable isomers **5a, b** are formed. Upon raising the temperature, **5a, b** rearrange into more thermodynamically stable **5c, d**. Initial formation of **5a, b** shows that the binding proceeds through the formation of the nonchelating complex **6a** via the coordination of the oxygen atom from the less hindered side of the catalyst.[12]

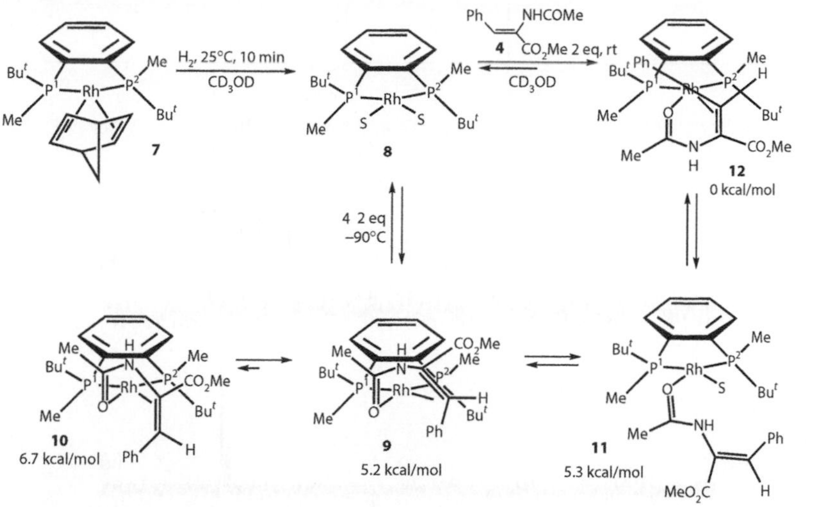

Scheme 1.3 Formation of catalyst–substrate complexes with catalyst **8** and substrate **4**. (Reprinted with permission from Imamoto, T. et al., *J. Am. Chem. Soc.*, 134, 1754–1769. Copyright 2012 American Chemical Society.)

Although the nonchelating complexes similar to **6** are unstable and were never detected directly, their kinetic availability can be demonstrated by NMR experiments testifying the possibility of intramolecular exchange between the chelating catalyst–substrate complexes.[13–15]

Another important intermediate was detected in a low-temperature reaction of substrate **4** with BenzP* catalyst **8** (Scheme 1.3).[16] Initially, a catalyst–substrate complex **9** with unusual *si-gauche* coordination of the double bond was formed. This was proved by reproducing computationally the specific ^{13}C chemical shifts of the coordinated double bond in **9** that are significantly different from those in *si-* and *re*-coordinated complexes **10** and **12** with normal orthogonal orientation of the double bond (Figure 1.1).[16]

Upon increasing the temperature, complex **9** disappeared from the equilibrium and a more thermodynamically stable *re*-coordinated complex **12** was observed instead. Figure 1.2 illustrates the difference in the coordination mode of the double bond in **9** and in **10, 12**. In **10** and **12** the double bond is orthogonal to the chelate cycle, and the α-carbon atom lies in the P–Rh–P plane. In **9** the double bond slides into the less hindered quadrant and becomes almost coplanar to the chelate cycle with the β-carbon atom lying in the P–Rh–P plane. This type of coordination is impossible for the *re*-coordinated double bond.[16]

It is also worth noting that both substituents at the double bond (phenyl and methoxycarbonyl) formally occupy the hindered quadrants in the most thermodynamically stable isomer **12**. This fact demonstrates the limited value of qualitative speculations on the relative stabilities of intermediates.

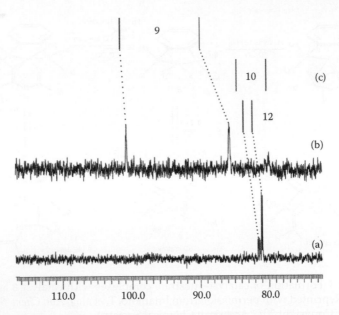

Figure 1.1 Section plots of ^{13}C NMR spectra (100 MHz, CD_3OD) showing signals from carbon atoms of the coordinated double bond. (a) Solution obtained by the addition of two equations of MAC to a deuteriomethanol solution of **7** at 25°C. Spectrum measured at 25°C. (b) The same reagents were mixed at –100°C and the sample was put into the pre-cooled probe of the NMR spectrometer. Spectrum measured at –90°C. (c) Computed chemical shifts (BLYP/SDD(Rh)6-31 + G(2d,2p)/CPCM(methanol)). (Reprinted with permission from Imamoto, T. et al., *J. Am. Chem. Soc.*, 134, 1754–1769. Copyright 2012 American Chemical Society.)

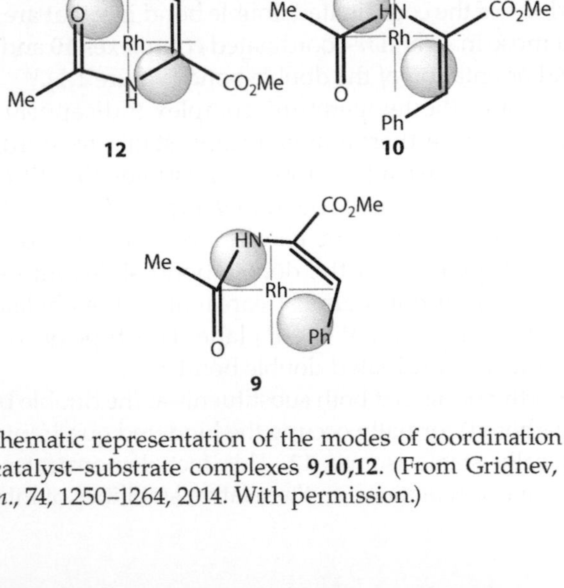

Figure 1.2 Schematic representation of the modes of coordination of the double bond in the catalyst–substrate complexes **9,10,12**. (From Gridnev, I. D., *J. Synth. Org. Chem. Jpn.*, 74, 1250–1264, 2014. With permission.)

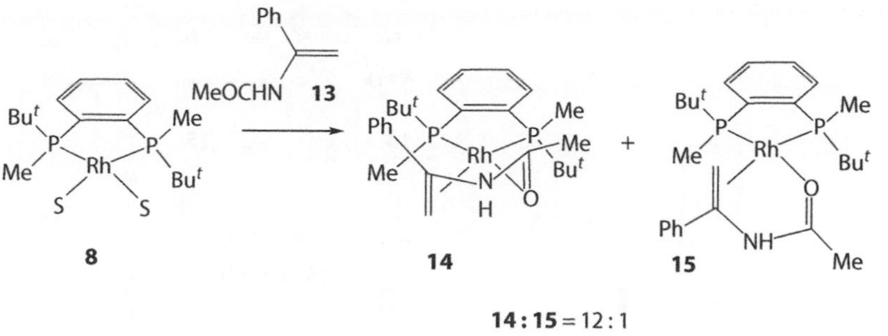

Scheme 1.4 Formation of the catalyst–substrate complexes **14** and **15**.

The reason for the greater thermodynamic stability of **12** can be better understood when analyzing the structures of the complexes of **8** with enamide **13**. In this case complex **14** with the phenyl ring adjacent to the *t*-Bu group of the catalyst was the most thermodynamically stable one (Scheme 1.4) that was confirmed both experimentally and computationally.[18]

The most illustrative evidence is the dramatic high-field shift of one of the *t*-Bu groups of the catalyst that is evidently due to its close proximity to the plane of the phenyl ring (Figure 1.3).

Apparently, instead of intuitively imaginable steric hindrance between the phenyl ring and *t*-Bu group, there is a kind of attraction that is most probably explained by C–H … π interaction. Similarly, the relative stability of complex **12** can also be explained.

Interestingly, the only catalyst–substrate complex that was observed at low temperature when **8** was reacted with the *t*-Bu-substituted enamide **16** was *re*-coordinated complex **17** in which the *t*-Bu groups from the catalyst and the substrate are positioned in the same quadrant (Scheme 1.5). Apparently, even extremely weak C–H … C–H interaction can facilitate the formation of the chelate cycle.[18]

Chelating catalyst–substrate complexes interconvert in solution at sufficiently high temperatures. This exchange can be either intermolecular (with complete dissociation of the substrate) or intramolecular (when only double bond dissociates). Most conveniently, these two mechanisms of exchange can be distinguished via 2D EXSY NMR experiments.[19]

Stability of the catalyst–substrate complexes relative to the mixture of a catalyst and a substrate varies in a broad range. The same is true for the relative stabilities of *si*- and *re*-coordinated species. For example, the position of the equilibrium between catalyst **8** and various β-dehydroamino acids changes from the complete absence of any binding in the case of (*E*)-3-acetylamino-2-butenoate **26** to only weak specific binding at low temperature with the corresponding (*Z*)-isomer **18** and almost complete binding at ambient temperature in the case of substrate **21**. The ratio of the *re*- and *si*-bound isomers also varies (Scheme 1.6).[20]

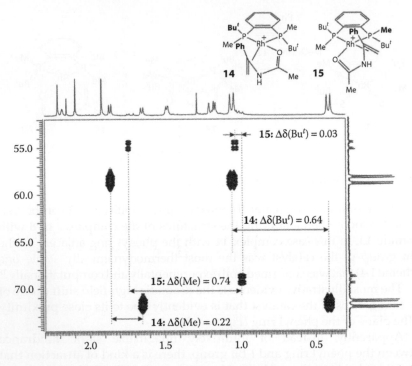

Figure 1.3 ¹H–³¹P HMBC NMR spectrum (600 MHz, CD₃OD, −60°C) of the equilibrium mixture of catalyst–substrate complexes **14** and **15** (section plot). Shielding of one of the *t*-butyl groups in **14** and one of the methyl groups in **15** defines the mode of the double bond coordination in both complexes, whereas the relative stabilities of these complexes suggest that a C–H…π interaction is the main stereoregulating factor. (Reprinted with permission from Gridnev, I. D. and Imamoto, T. *ACS Catal.*, 5, 2911–2915. Copyright 2015 American Chemical Society.)

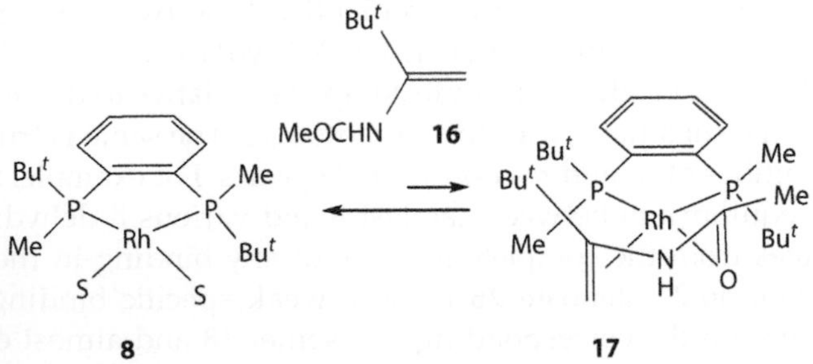

Scheme 1.5 Formation of the catalyst–substrate complex **17**.

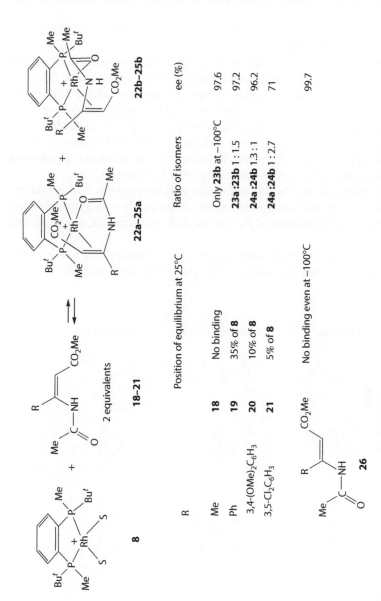

R		Position of equilibrium at 25°C	Ratio of isomers	ee (%)
Me	**18**	No binding	Only **23b** at −100°C	97.6
Ph	**19**	35% of **8**	**23a**:**23b** 1 : 1.5	97.2
3,4-(OMe)$_2$C$_6$H$_3$	**20**	10% of **8**	**24a**:**24b** 1.3 : 1	96.2
3,5-Cl$_2$C$_6$H$_3$	**21**	5% of **8**	**24a**:**24b** 1 : 2.7	71
	26	No binding even at −100°C		99.7

Scheme 1.6 Formation of catalyst–substrate complexes of catalyst **8** and esters of β-dehydroamino acids. (From Gridnev, I. D. et al., *ACS Catal.*, 4, 203–219, 2014. With permission.)

From the data shown in Scheme 1.6, it is clear that the absence of the substrate binding does not prevent high enantioselectivity; moreover the higher optical yields are observed in the cases with weak or absent binding.

1.1.2.2   Activation of H_2

It was shown in the previous section that the reaction pool of the Rh-catalyzed asymmetric hydrogenation before the admission of dihydrogen contains numerous equilibrating species, that is, solvate complexes, nonchelating catalyst–substrate complexes, and chelating catalyst–substrate complexes with different coordination modes of the prochiral double bond. Hence, there is a possibility of three different modes of dihydrogen activation: via solvate complexes **2**, via nonchelating catalyst–substrate complexes **4**, and via chelating catalyst–substrate complexes **5** (Scheme 1.7).

Which pathway for the H_2 activation will be prevailing in a catalytic system depends mostly on the structures of the catalyst and the substrate, but this can be also influenced by the solvent, temperature, pressure, and so on.

Table 1.1 shows thermodynamic parameters characterizing relative affinities of different solvate complexes to H_2. Although these affinities can vary

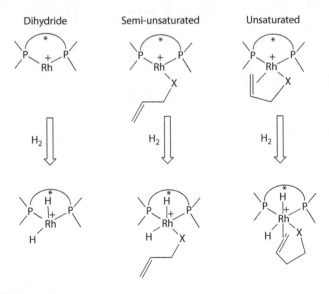

Scheme 1.7 Three possible modes of hydrogen activation in Rh-catalyzed asymmetric hydrogenation of activated olefins. For the clarity reasons here and further the coordinated molecules of solvent completing square planar for Rh(I) or octahedral for Rh(III) configuration are not shown.

Table 1.1 Gibbs free energies (298 K, kcal/mol) characterizing equilibria between solvate complexes, molecular hydrogen complexes, and solvate dihydride complexes for various Rh-diphosphine catyalysts computed at WB97XD/ SDD(Rh)/6-31G(d, p)(all others)/SMD(methanol) level of theory

Entry	Catalyst	ΔG^1	ΔG^2	ΔG^3	$\Delta G^{\neq 1}$	$\Delta G^{\neq 2}$	$\Delta G^{\neq 1}{}_{eff}$[a]	$\Delta G^{\neq 2}{}_{eff}$[b]
1	(S, S)-DIPAMP	15.9	−4.9	−5.7	2.0	2.0	17.9	17.9
2	(R, R)-BenzP*	10.0	−2.8	−3.1	3.4	3.0	13.4	13.0
3	(S)-BINAP	4.4	−1.2	0.5	4.4	3.8	8.8	8.2
4	(S, S′,R, R′)-TangPhos	1.6	−2.4	−2.8	7.0	4.3	8.6	5.9

(Continued)

Table 1.1 (Continued) Gibbs free energies (298 K, kcal/mol) characterizing equilibria between solvate complexes, molecular hydrogen complexes, and solvate dihydride complexes for various Rh-diphosphine catyalysts computed at WB97XD/ SDD(Rh)/6-31G(d, p)(all others)/SMD(methanol) level of theory

Entry	Catalyst	ΔG^1	ΔG^2	ΔG^3	$\Delta G^{\neq 1}$	$\Delta G^{\neq 2}$	$\Delta G^{\neq 1}_{eff}$ [a]	$\Delta G^{\neq 2}_{eff}$ [b]
5	(R, R)-QuinoxP*	1.4	−2.6	−1.1	7.4	6.9	8.8	8.3
6	(R, R)-DioxyBenzP*	1.1	−1.5	−3.2	4.6	2.4	5.6	3.5
7	(S, S)-CHIRAPHOS	0.8	−2.0	−3.1	0.6	4.2	1.4	5.0
8	(S, S)-BisP*	−0.1	−3.0	−3.8	3.8	3.8	3.7	3.7

(Continued)

Table 1.1 (Continued) Gibbs free energies (298 K, kcal/mol) characterizing equilibria between solvate complexes, molecular hydrogen complexes, and solvate dihydride complexes for various Rh-diphosphine catyalysts computed at WB97XD/SDD(Rh)/6-31G(d, p)(all others)/SMD(methanol) level of theory

Entry	Catalyst	ΔG^1	ΔG^2	ΔG^3	$\Delta G^{\neq 1}$	$\Delta G^{\neq 2}$	$\Delta G^{\neq 1}_{eff}$[a]	$\Delta G^{\neq 2}_{eff}$[b]
9	(R, R)-Me-BPE	−0.7	−2.0	−2.6	3.2	3.8	1.5	2.1
10	(R, R)-Me-DuPHOS	−1.2	−4.3	−3.8	3.7	4.8	2.5	3.6
11	(R)-Trichickenfootphos	−1.5[c]	−1.3	−2.3	4.7	4.7	3.2	3.2
		−1.9[d]	−1.7	−2.2	5.1	4.8	3.2	2.9

Source: Gridnev, I. D. and Imamoto, T. *Izv. Akad. Nauk, Ser. Khim.*, 1514–1534, 2016.

[a] $\Delta G^{\neq 1}_{eff} = \Delta G^1 + \Delta G^{\neq 1}$.
[b] $\Delta G^{\neq 2}_{eff} = \Delta G^1 + \Delta G^{\neq 2}$.
[c] H_2 nearby the nonchiral phosphorus atom.
[d] H_2 nearby the chiral phosphorus atom.[21]

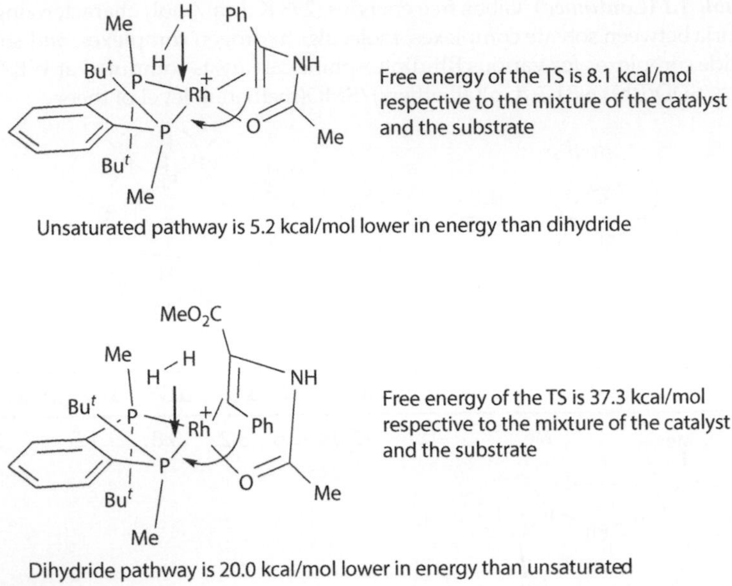

Free energy of the TS is 8.1 kcal/mol respective to the mixture of the catalyst and the substrate

Unsaturated pathway is 5.2 kcal/mol lower in energy than dihydride

Free energy of the TS is 37.3 kcal/mol respective to the mixture of the catalyst and the substrate

Dihydride pathway is 20.0 kcal/mol lower in energy than unsaturated

Scheme 1.8 Influence of the substrate structure on the mechanism of H_2 activation.

in some range, one can see that for the most of the catalysts the solvate dihydrides are kinetically available. Only for DIPAMP-Rh (entry 1) the hydrogen activation via solvate complexes lacking additional complexation seems to be hardly probable due to high activation barrier. But already in the case of the BenzP*-Rh catalyst (entry 2), it depends on the nature of the substrate.

This is illustrated in Scheme 1.8. Hydrogenation of BenzP*-Rh catalyst 8 is characterized by relatively high activation barrier (Table 1.1, entry 2), and in the case of the substrate with unsubstituted β-carbon atom, the activation of dihydrogen proceeds via unsaturated mechanism with lower activation barrier.[18] Nevertheless, with the substrate having β-substituent, the activation barrier for oxidative addition to a chelating catalyst–substrate complex becomes very high, and H_2 activation proceeds via dihydride mechanism.[22]

1.1.2.3 Formation of solvate dihydrides

Experimentally formation of solvate dihydrides has been observed for several Rh–diphosphine catalysts. Thus, two diastereomeric dihydrides **28a, b** were obtained via the reversible oxidative addition of H_2 to the solvate complex [Rh(*t*-Bu-BisP*)(CD$_3$OD)$_2$]$^+$ (**27**) and have been characterized by NMR (Scheme 1.9).[23]

By comparing the results of ^{1}H–^{1}H and ^{31}P–^{31}P 2D EXSY spectra, it was possible to conclude that **28a** and **28b** can interconvert without the

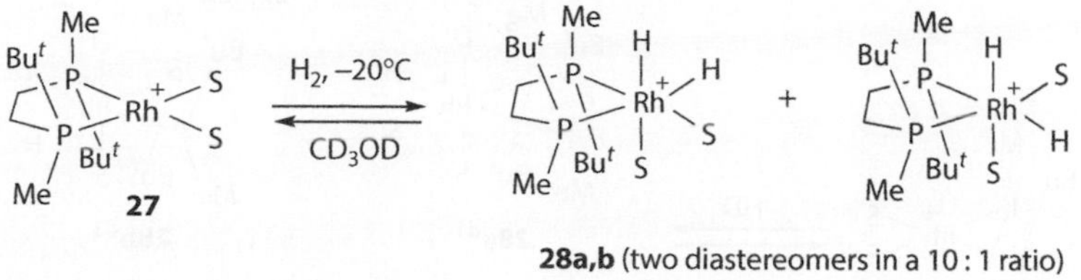

28a,b (two diastereomers in a 10 : 1 ratio)

Scheme 1.9 Experimental detection of solvate dihydrides **28**. (Reprinted with permission from Gridnev, I. D. et al., *J. Am. Chem. Soc.*, 122, 7183–7194. Copyright 2000 American Chemical Society.)

28a **29** **28b**

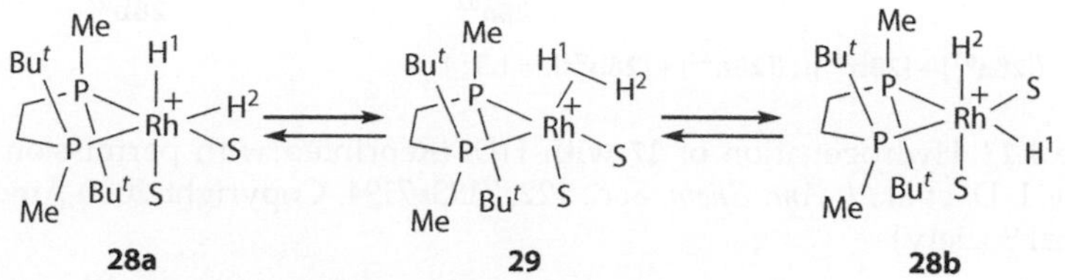

Scheme 1.10 Interconversion of **3a, b** via molecular hydrogen complex **7**. (Reprinted with permission from Gridnev, I. D. et al., *J. Am. Chem. Soc.*, 122, 7183–7194. Copyright 2000 American Chemical Society.)

dissociation of dihydrogen. This proved the intermediacy of the molecular hydrogen complex **29** (Scheme 1.10).[23]

When HD was used for the generation of the solvate dihydrides **28**, the complexes with axial position of the deuterium atom were formed preferentially with the factor 1.3:1 (Scheme 1.11).[23] These data were later used for tracking the position of the deuterium label in the hydrogenation product (*vide infra*).

Similar observations of reversible formation of solvate dihydrides were made for other BisP*–Rh complexes,[24] for the TangPhos–Rh complex,[22] and for the PHANEPhos–Rh complex.[25,26]

On the other hand, the irreversible formation of the two diastereomeric dihydrides **31a, b** was observed when the catalytic precursor having an *o*-xylyl backbone (**30**) was hydrogenated.[27] Similarly, the stereoselective irreversible formation of a tetrahydride species **33** was observed in the hydrogenation of the binuclear tetraphosphine–Rh complex **32**[28] (Scheme 1.12). No intermediate solvate complexes were detected in these two hydrogenation experiments.

The examples of direct observation of solvate dihydrides together with the computational data collected in the Table 1.1 illustrate the fact that these species are kinetically competent intermediates in the catalytic cycle of Rh-catalyzed asymmetric hydrogenation.

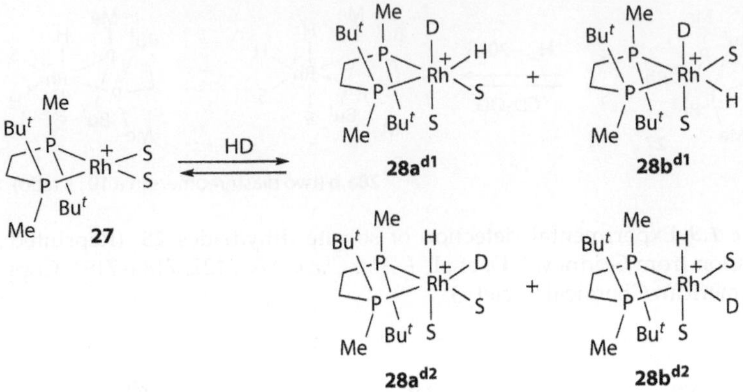

$$([28a^{d1}] + [28b^{d1}]) : ([28a^{d2}] + [28b^{d2}]) = 1.3 : 1$$

Scheme 1.11 Hydrogenation of **27** with HD. (Reprinted with permission from Gridnev, I. D. et al., *J. Am. Chem. Soc.*, 122, 7183–7194. Copyright 2000 American Chemical Society.)

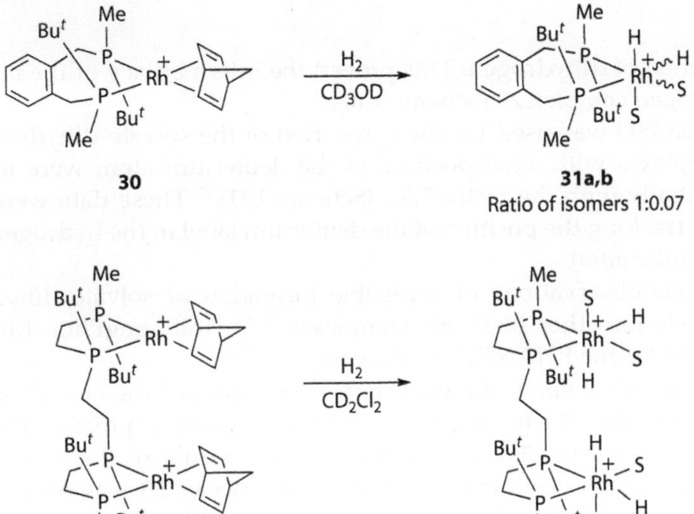

30

31a,b
Ratio of isomers 1:0.07

32

33

Scheme 1.12 Irreversible formation of solvate dihydrides. (Gridnev, I. D. and Imamoto, T., *Chem. Commun.*, 7447–7464, 2009. Reproduced by permission of The Royal Society of Chemistry.)

1.1.2.4 *Oxidative addition to chelate catalyst–substrate complexes*

Since the chelating catalyst–substrate complexes with different mode of double bond coordination are diastereomers, they have different chemical properties, in particular, the rates of oxidative addition must be different. Moreover, it is clear that hydrogen atoms must come from the side of rhodium. Hence, if the oxidative addition is irreversible and the double bond does not dissociate before, during, or after the oxidative addition, the original mode of the double bond coordination could determine the stereochemical outcome of the reaction via the difference in reactivities of the corresponding catalyst–substrate complexes. This is the main idea of the so-called unsaturated mechanism of stereoselection (Scheme 1.13).[29,30]

Although totally eight possible pathways for oxidative addition could be conceivable for any combination of a chiral Rh complex and a prochiral substrate, computations show that only three of them have reasonable activation barriers due to electronic factors or hindrance from the substituents on phosphorus atoms and coordinated substrate.[31-34] Either *re*-coordinated catalyst–substrate complex **34** or *si*-coordinated complex **35** can undergo oxidative addition via coordination of dihydrogen over Rh–P bond in *trans*-position respective to the coordinated double bond (Scheme 1.12). Besides, *si*-coordinated complex **35** can rearrange to *si-gauche*-coordinated complex **36** by shifting the chelate cycle to almost axial position in the less hindered quadrant. That opens a possibility for an alternative pathway of dihydrogen approach. If the double bond would keep its coordination during the oxidative addition (leading to the corresponding dihydride complexes **40–42**), and migratory insertion step leading to monohydrides **43–45** would be irreversible, then the optical yield would be determined via the relative rates of the H_2 oxidative addition to **34–36** (Scheme 1.13).

At the early stage of the mechanistic studies of Rh-catalyzed asymmetric hydrogenation, this mechanism has been accepted on the basis of low-temperature hydrogenation experiments clearly demonstrating higher reactivity of the catalyst–substrate complexes with "proper" mode of coordination of the double bond toward H_2.[35,36]

However, more recent experiments showed that the whole catalytic cycle until the migratory insertion stage can be reversible, and the double bond can dissociate before, during, or after the oxidative addition, thus deleting any chiral information acquired so far.

Thus, Brown et al. studied the low-temperature reaction of the solvate dihydride **47** with MAC (**4**) and the hydrogenation of an equilibrium mixture of the solvate **48** and the catalyst–substrate complex **49** (Scheme 1.14).[25,26] Both experimental approaches afforded the same intermediate, agostic complex **50** (apparently through intermediates **51** and **52**). The spectroscopic evidence clearly showed that one of the hydrides in **50** is caught on its way from Rh to carbon in this intermediate,

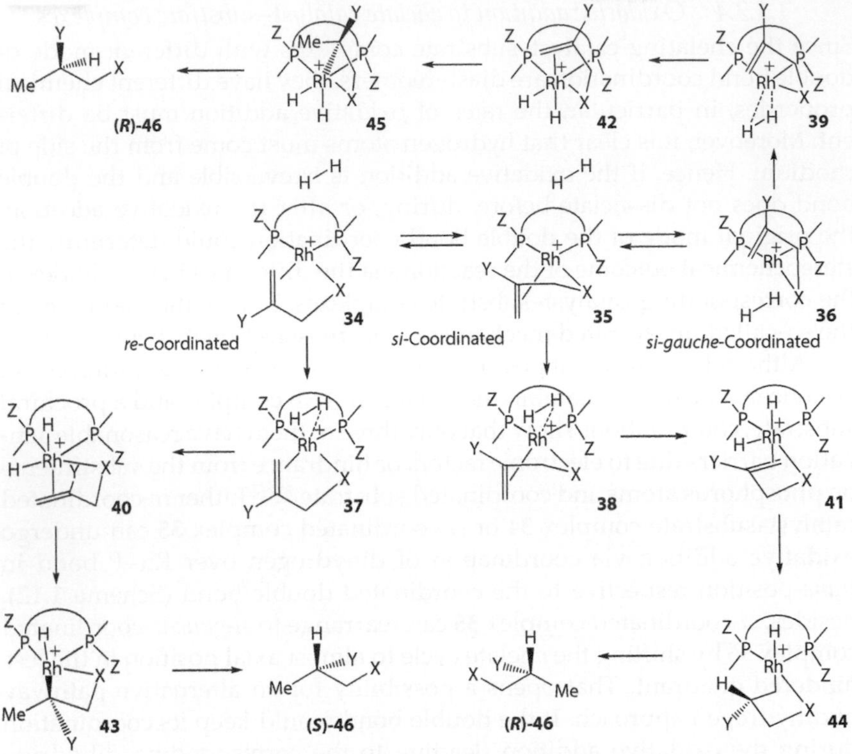

Scheme 1.13 Stereoselection via unsaturated pathway.

whereas the remaining hydride is *trans* to the oxygen atom. According to the computational results, an intermediate of such structure (double bond *trans* to phosphorus) is not accessible through the direct oxidative addition of H_2 to **49**, hence the hydrogenation of **49** has to proceed via the initial dissociation of the double bond (yielding **53**) or the entire substrate molecule (returning to **48** that is further hydrogenated to **47**).

Importantly, the formation of **50** was stereoselective. On the other hand, **50** was shown to exist in an equilibrium with the catalyst–substrate complex **49**, the solvate dihydride **47** along with the mixture of catalyst **48** and MAC.[26] Stereochemistry of the hydrogenation product is completely defined in **50**, but the reverse reaction can bring it back to numerous interconverting intermediates with flexible stereochemistry. Nevertheless, when **50** is formed again, it results in exactly the same structure that clearly defines the absolute configuration of the hydrogenation product.

Low-temperature hydrogenation experiments of catalyst–substrate complexes **5a–c** demonstrated that hydrogenation is accompanied by dissociation of the double bond (Scheme 1.15).[12] The ee of the product obtained

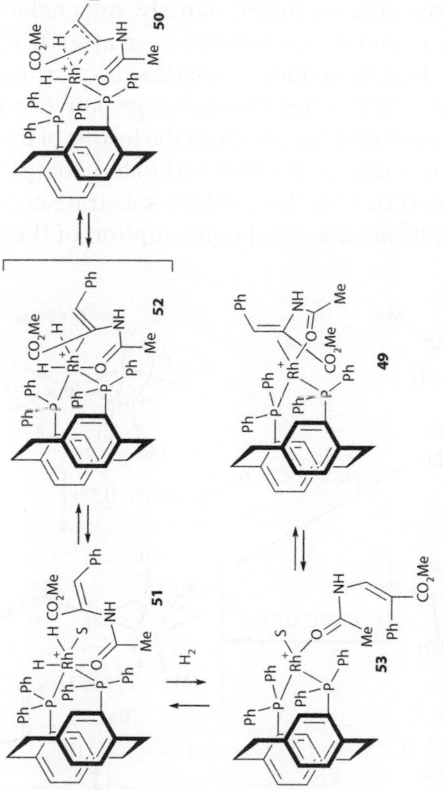

Scheme 1.14 Reversible formation of the agostic intermediate **50**. (Adapted from Heinrich, H. et al. *Chem. Commun.*, 1296–1297, 2001. With permission.)

after the complete hydrogenation of the mixture of **5c, d** under conditions precluding their interconversion was 97% (*R*)—exactly the same ee was obtained when the sample was quenched after only the **5d** has been consumed. When the kinetic product **5a** was hydrogenated before it could rearrange to **5c, d**, again the ee of the hydrogenation product was 97% (*R*).[12] These results require that hydrogenation of either of **5a–d** should occur through common intermediates, namely, nonchelating complexes **6a, b** that are capable of fast interconversion even at −100°C.

Low-temperature hydrogenation experiments for the equilibrium mixture of **14** and **15** at −78°C when the exchange between these catalyst–substrate complexes is slow and cannot affect the results of the reaction demonstrated more rapid reaction of the *re*-coordinated complex **14**.[20] Similar experiments were carried out for the catalyst–substrate complexes **55** and **56** (Scheme 1.16).[21] In all cases, a rapid consumption of the *re*-coordinated

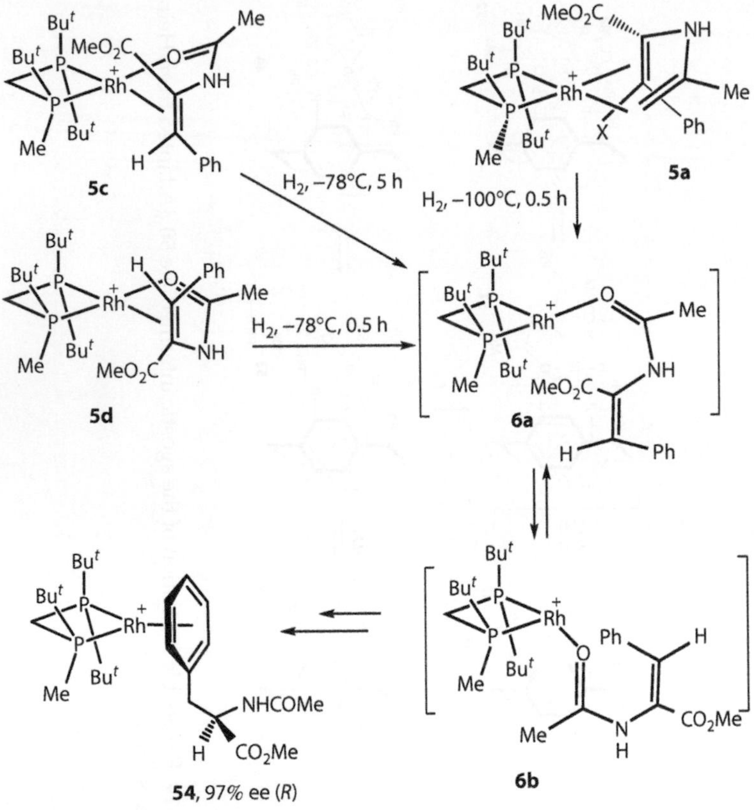

Scheme 1.15 Three low-temperature hydrogenation experiments yielding the same result. (Gridnev, I. D. and Imamoto, T., *Chem. Commun.*, 7447–7464, 2009. Reproduced by permission of The Royal Society of Chemistry.)

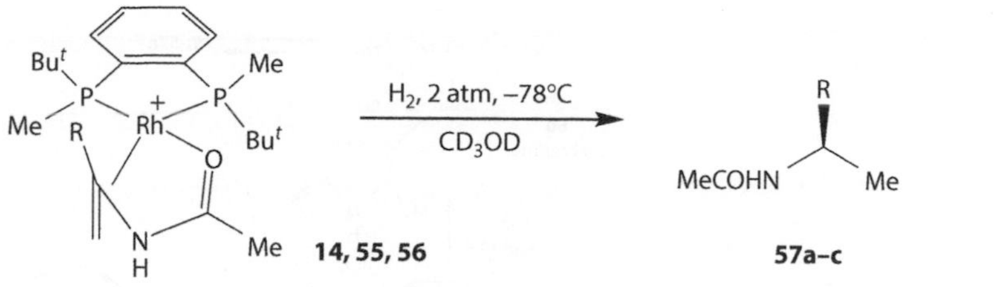

14 (+15)	R = C_6H_5	98.9% ee (*R*), 98.1% ee (*R*)
55	R = *p*-(NO_2)C_6H_4	99.2% ee (*R*)
56	R = 3,5-(CF_3)$_2C_6H_3$	98.6% ee (*R*)

Scheme 1.16 Low-temperature hydrogenation experiments proving dissociation of the double bond after oxidative addition of H_2.

complexes was observed to afford directly the *R*-hydrogenation products. Hence, these experiments demonstrated explicitly that the observation of a rapid reaction with a catalyst–substrate complex does not necessarily mean that the handedness of the product would be determined by the mode of the double bond coordination in this catalyst–substrate complex.

Computations showed that indeed in that case the oxidative addition of H_2 to the *re*-coordinated complex, for example, **14**, is the fastest mechanism of dihydrogen activation among all conceivable processes (hydrogenations of solvate complex, of nonchelating catalyst–substrate complexes and of *si*-coordinated catalyst–substrate complexes are slower). Nevertheless, after formation of the dihydride intermediate **58**, the double bond dissociation via the **TS3** is much faster than migratory insertion via the **TS2** (Scheme 1.17). As a result, the nonchelating dihydride intermediate **60** is formed, and (*R*)-**57a** is formed after stereoselective double bond coordination in a Rh(III) octahedral nonchelating complex (*vide infra*).[20,21]

These results prompted a reconsideration of the interpretation of early experiments demonstrating selective hydrogenation of one chelating catalyst–substrate complex.[35,36] Since the double bond can dissociate before or after the oxidative addition step, the observation of the selectivity on this stage of the reaction does not automatically mean that the handedness of the product is already determined and fixed.

Thus, the computations involving real molecules of the Rh-DIPAMP catalyst and MAC used in the famous experiment showed that direct oxidative addition is hardly possible to either of the experimentally observed complexes **61** and **62**. On the other hand, the *si*-coordinated catalyst–substrate complex **62** is capable of facile rearrangement to a *si-gauche*-coordinated complex **63** that has not been observed experimentally due to its relative instability (Scheme 1.18).[21]

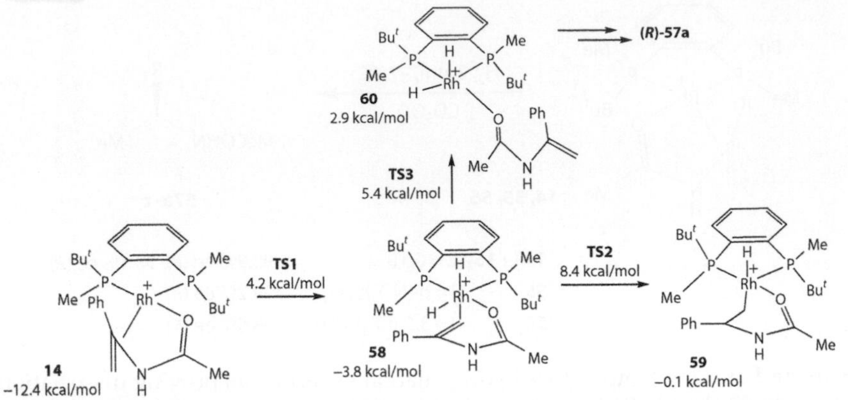

Scheme 1.17 Competition between migratory insertion and double bond dissociation in the dihydride intermediate **58**. (Reprinted with permission from Gridnev, I. D. and Imamoto, T. *ACS Catal.*, 5, 2911–2915. Copyright 2015 American Chemical Society.)

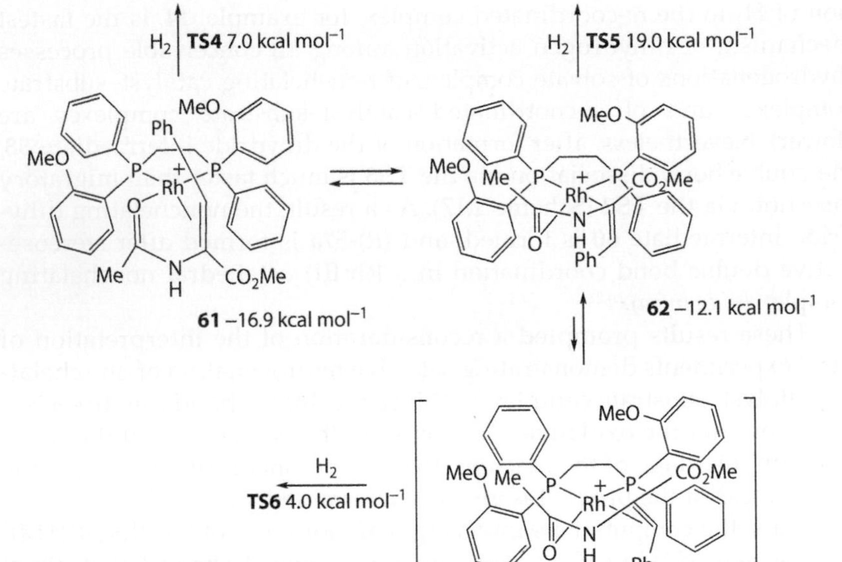

Scheme 1.18 Oxidative addition of H$_2$ to the catalyst–substrate complexes **61–63**. (From Gridnev, I. D. and Imamoto, T. *Izv. Akad. Nauk, Ser. Khim.*, 1514–1534, 2016. With permission.)

Thus, the experimentally observed rapid consumption of **62** upon low-temperature hydrogenation of the mixture of **61** and **62** took place not directly, but via isomerization of **62** to another, much less stable, but significantly more reactive species **63**.

Further computations showed that the dihydride intermediate **64** formed after oxidative addition of hydrogen to **63** is capable of facile migratory insertion resulting in the monohydride intermediate **65** (Scheme 1.19). However, the *trans*-configuration of the hydride and alkyl ligands in **65** prevents the direct reductive elimination, whereas the rearrangement into more reactive monohydride requires significant activation energy, and the dissociation of the double bond via reverse migratory

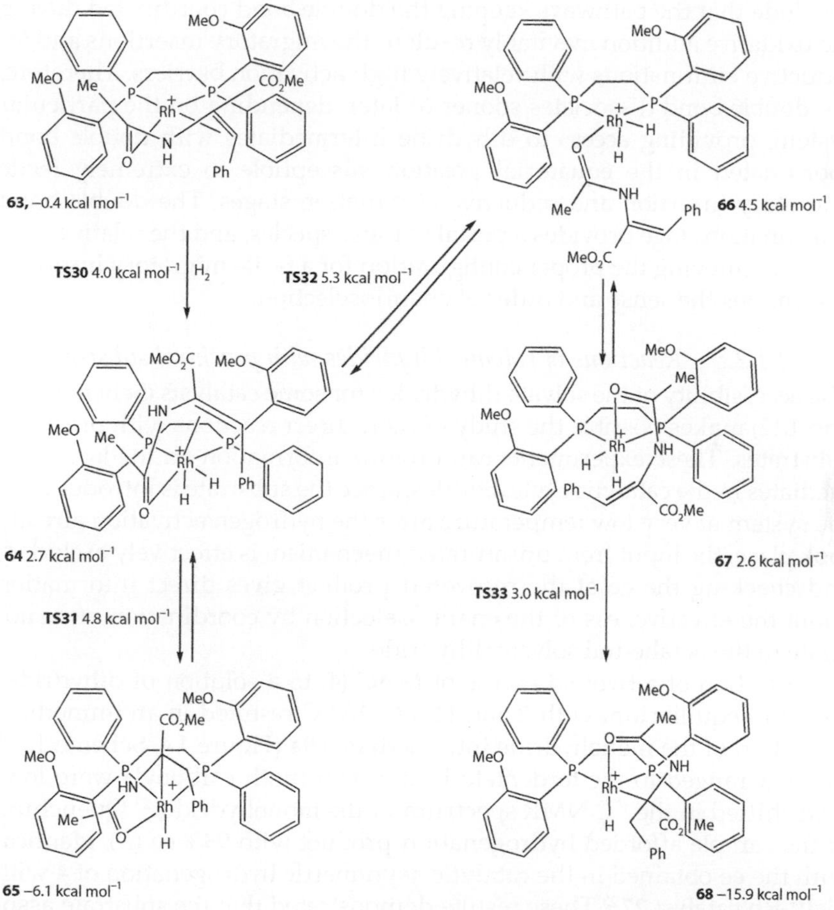

Scheme 1.19 Competition between migratory insertion-reductive elimination and double bond dissociation in the dihydride intermediate **64**. (From Gridnev, I. D. and Imamoto, T. *Izv. Akad. Nauk, Ser. Khim.*, 1514–1534, 2016. With permission.)

insertion yielding nonchelating Rh(III) complex **66** is a much more facile process (Scheme 1.19).[21] Coordination of the double bond followed by facile migratory insertion yields monohydride **68** which is 9.8 kcal/mol more stable than **65** and was most probably observed experimentally.

Thus, although in that case the coordination of the double bond can be kept until the very late stage of the catalytic cycle, the reductive elimination is not possible within this pathway, and again the trail involving dissociation of the double bond must be taken. This leads to the loss of any chiral information acquired so far, and enantioselection must take place during the double bond coordination in the nonchelating Rh(III) dihydride intermediate **66**.

Thus, the analysis of the available experimental and computational data for broad range of the combinations catalyst–substrate allows us to conclude that the pathways keeping the double bond coordinated during the oxidative addition inevitably result in the migratory insertions and/or reductive eliminations with relatively high activation barriers. Therefore, the double bond dissociates sooner or later, depending on the particular system, providing access to dihydride intermediates with double bond coordinated in the equatorial position susceptible to extremely facile migratory insertion and reductive elimination stages. The double bond coordination stage provides a portal to these species, and the relative easiness in achieving the proper configuration for a facile migratory insertion determines the sense and order of enantioselection.

1.1.2.5   Reactions of solvate dihydrides with prochiral substrates

The accessibility of the solvate dihydrides for some catalysts (Schemes 1.10 and 1.12) makes possible the study of their direct reactions with prochiral substrates. These experiments can provide information on the late intermediates in the catalytic cycle. Besides, since the substrate is introduced in the system at very low temperature after the hydrogen activation already took place, the input from unsaturated mechanism is effectively excluded, and checking the ee of the recovered product gives direct information about the effectiveness of the enantioselection by coordination of a substrate to the octahedral solvate dihydride.

Addition of a twofold excess of MAC (**4**) to a solution of dihydrides **28a, b** in equilibrium with **2** and H_2 at −100°C resulted in an immediate formation of the monohydride intermediate **69a** (Figure 1.4, Scheme 1.20) that rearranged to the terdentate isomer **69b** (both carbonyls were low-field shifted in the ^{13}C NMR spectrum of the monohydride).[23] Quenching of the sample afforded hydrogenation product with 99% ee (*R*), identical with the ee obtained in the catalytic asymmetric hydrogenation of **4** with BisP*-Rh catalyst **27**.[23] These results demonstrated that the substrate association yielding nonchelating complex **70**, double bond coordination providing dihydride intermediate **71**, and the subsequent migratory insertion step proceed extremely fast and highly enantioselectively.

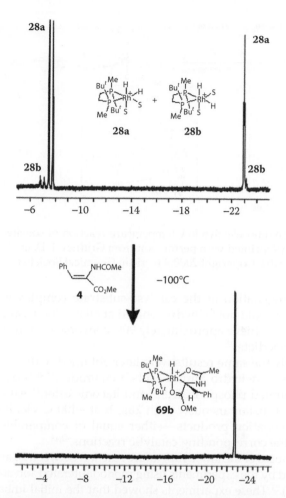

Figure 1.4 Hydride region of ^{1}H NMR spectra (600 MHz, CD$_3$OD, –95°C) of the sample initially containing solvate dihydrides **28a, b** in equilibrium with the solvate complex **27** at –100°C obtained by placing the sample to the probe precooled to –100°C. Immediate stereoselective reaction takes place yielding the corresponding monohydride intermediate. The fact that the equatorial hydride *trans* to phosphorus (δ = –7.7, $^2J_{H-P}$ = 186 Hz) was transferred in this reaction is well illustrated. (Gridnev, I. D. and Imamoto, T., *Chem. Commun.*, 7447–7464, 2009. Reproduced by permission of The Royal Society of Chemistry.)

The association of substrate **4** begins by coordinating the oxygen atom to **27**, affording the nonchelating dihydride **70**. Coordination of the double bond occurs to form **71** in which the chelate cycle is in the less hindered quadrant and the double bond is parallel to the Rh–H^{trans} bond. Immediately after the formation of **71**, migratory insertion takes place yielding **6**, and the hydrogenation product **(R)-72** after quenching.[23]

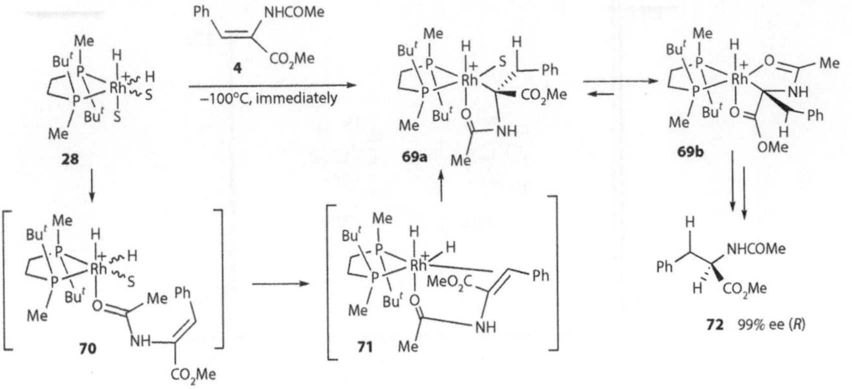

Scheme 1.20 Enantioselective low-temperature reaction of solvate dihydride **28** with MAC (**4**). (Reprinted with permission from Gridnev, I. D. et al., *J. Am. Chem. Soc.*, 122, 7183–7194. Copyright 2000 American Chemical Society.)

The hydrogenation of the catalyst–substrate complex was substantially slower: it could not be hydrogenated at all at −100°C, and it required 2 h at −80°C to achieve approximately 90% conversion to the same monohydride intermediate.[23]

Essentially the same results were later obtained with various prochiral substrates: α-dehydro amino acids,[24] enamides,[37,38] β-dehydroamino acid,[39] unsaturated phosphonate,[40,41] and itaconic ester.[24] All studied substrates reacted instantaneously with **28a, b** at −100°C yielding high ee's of the hydrogenation products—either equal or comparable to the ee's observed in the corresponding catalytic reactions.[42,43]

The rapidness of the low-temperature reaction between **28** and **4** was further illustrated by the experiments using monodeuterated solvate dihydrides (Scheme 1.21).[23] These experiments showed that the initial imbalance in the deuterium distribution in favor of the axial position is roughly conserved within low-temperature reactions of the HD solvate dihydrides with the substrate and is equal to that observed under the catalytic conditions. This is reflected in the preferential formation of α-deuterated product. Therefore, the formation of the monohydride **69** via **71** must have taken place much faster than the scrambling of the positions of the hydride and the deuteride ligands (the latter has been measured as 1.4 s⁻¹ at −80°C for the H₂ case).[23]

The reaction of the solvate dihydride **28** with dimethyl 1-benzoyloxy-ethenephosphonate (**73**) at −100°C gave the hydrogenation product **74** with up to 97% ee (*S*) via the monohydride complex **75** (Scheme 1.22).[40,41] Compound **73** exhibits stronger chelate binding than **4**, and accordingly the hydrogenation of the catalyst–substrate complex (single isomer **76** with *si*-coordinated double bond has been observed) could only be performed at −30°C affording 75% ee (*S*). Similarly to the previous case, the

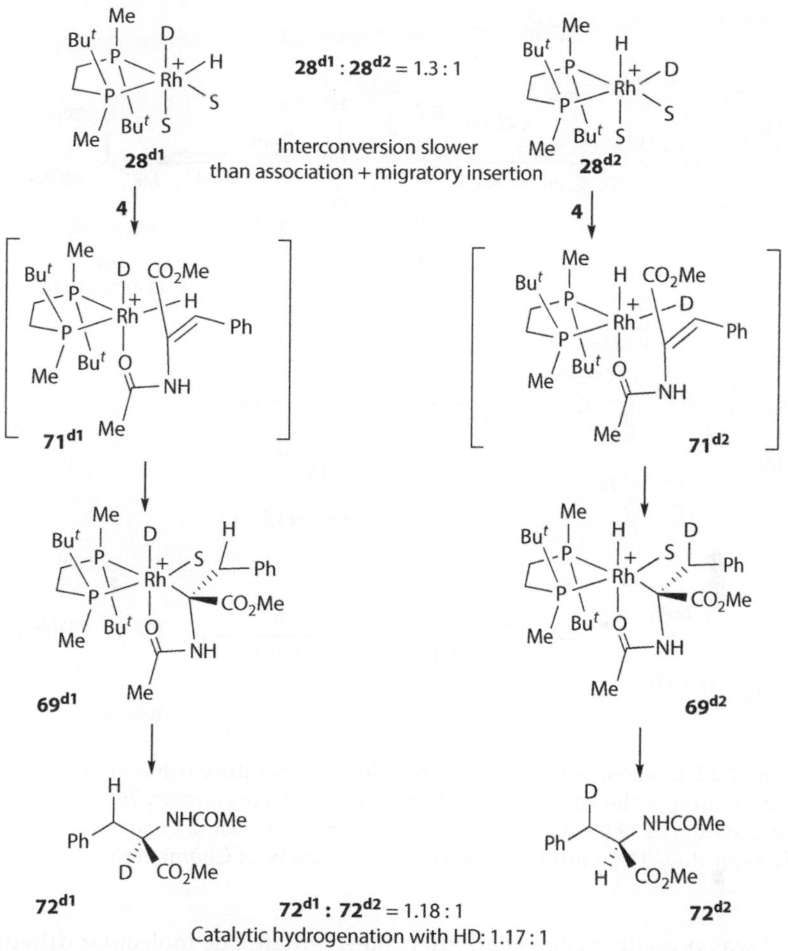

Scheme 1.21 Distribution of deuterium in the hydrogenation product in low-temperature experiments and in catalytic hydrogenation using HD. (Reprinted with permission from Gridnev, I. D. et al., *J. Am. Chem. Soc.*, 122, 7183–7194. Copyright 2000 American Chemical Society.)

coordination mode of the prochiral double bond in **76** did not correspond to the configuration of the hydrogenation product, hence the double bond must have dissociated before the enantiodetermining step. The catalytic reaction gave an intermediate value (88% ee) suggesting that under the catalytic conditions a certain amount of the strongly bound phosphonate is hydrogenated via the direct oxidative addition of H_2 to **76** decreasing the ee of the product.[41]

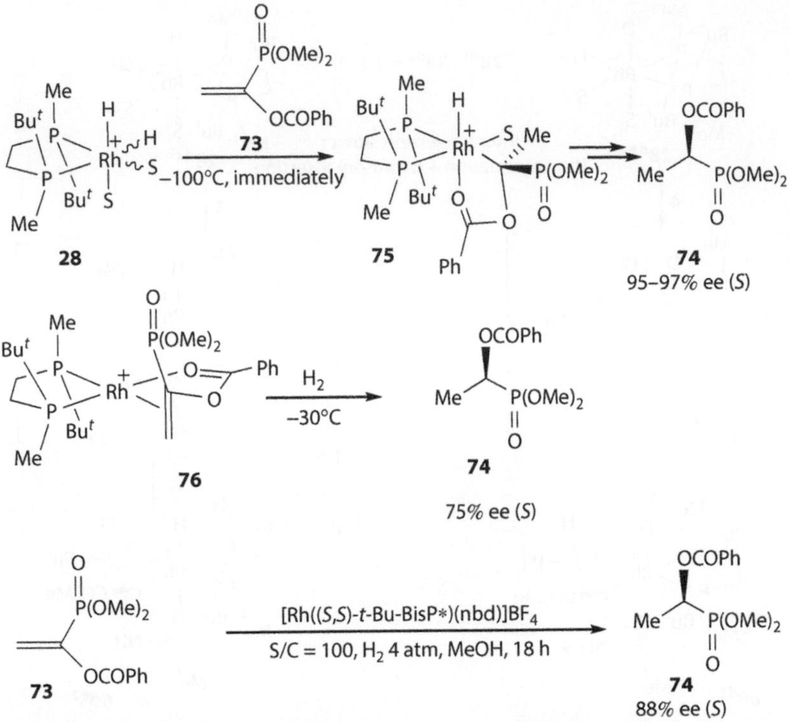

Scheme 1.22 Comparison of the results of low-temperature reaction of **28** and **73**, hydrogenation of the strongly bound catalyst–substrate complex **76**, and catalytic hydrogenation of **73**.[40,41] (Gridnev, I. D. and Imamoto, T., *Chem. Commun.*, 7447–7464, 2009. Reproduced by permission of The Royal Society of Chemistry.)

It was possible to detect another intermediate, the molecular dihydrogen complex **77**, when the phosphonate **73** ^{13}C-labeled α to the P atom was used in the low-temperature reaction with **28** (Scheme 1.23).[41] The position and the coupling pattern of the α-carbon atom, the high-field shift of one of the phosphorus atoms, as well as fast conversion of this species to the monohydride **75** at $-95°$C gave good reasons to assume structure **77**.[41]

Apparently, the substrate can react not only with the dihydrides **28a, b**, but also with the molecular hydrogen complex **29** which is equilibrating with **28a, b** and **27** (*vide supra*). That would yield nonchelating dihydride **78** and nonchelating molecular hydrogen complex **79**, respectively. The following association of the double bond would produce the dihydride intermediate **80** (which immediately yields **75**) and **77** which can be detected before it transforms into **80** either directly or via **78** and **79** (Scheme 1.23).[41]

Important results were obtained when the reaction of tetrahydride **33** with MAC (**4**) was studied.[28] Despite the fact that there were two solvate

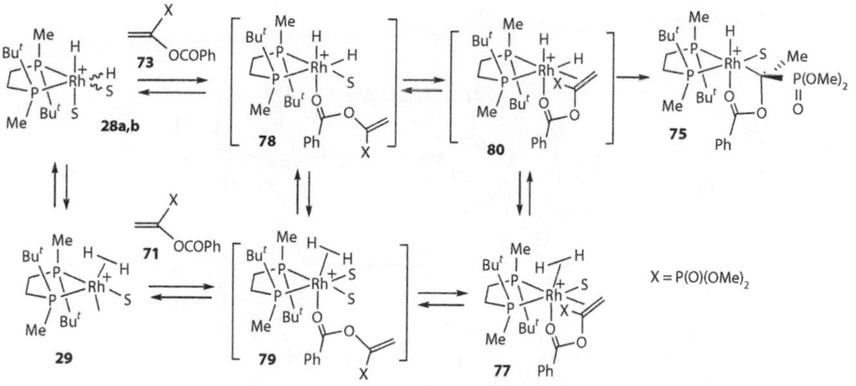

Scheme 1.23 Possible pathways leading to **75**. (From Gridnev, I. D. et al., *Proc. Natl. Acad. Sci. U.S.A.*, 101, 5385–5390, 2004. With permission.)

dihydride units, one of them remained intact even in the presence of the excess of **4**. The intermediate **81**, which formed quantitatively immediately after the preparation of the sample at −100°C, contained one molecule of MAC coordinated in a nonchelating manner (Scheme 1.24). This was confirmed by conducting the same experiment with β-^{13}C-labeled MAC.[28]

Although **81** contained in the coordination sphere both an activated hydrogen and the coordinated substrate, it did not undergo the migratory insertion itself. Instead, a degenerate rearrangement due to the exchange of the coordinated MAC between two equivalent coordination sites was observed at temperatures between −100°C and −50°C. At −20°C the rearrangement of **81** to its isomer **82** was accompanied by migratory insertion resulting in the trihydride **84** (Scheme 1.24). Quenching of the sample afforded the hydrogenation product **72** with 95% ee (*R*).[28]

Thus, in this experiment the quantitative formation of the nonchelating dihydride–substrate complexes **81** and **82** has been observed. Obviously, the coordination of the double bond with the formation of chelating tetrahydride–substrate complex **83** must occur before the migratory insertion. However, due to an essentially negligible barrier of the migratory insertion (less than 1 kcal/mol), chelating dihydride intermediate **83** was not observed directly.

Remarkable is the striking difference in the sense of enantioselection observed in the Rh-catalyzed asymmetric hydrogenation of α-phenyl (**13**) and α-*t*-butyl (**16**)-substituted enamides with Rh-DuPhos catalyst (Scheme 1.25).[30] Later, similar results were obtained with the use of Rh-*t*-BuBisP*, Rh-BenzP*, Rh-QuinoxP*, and Rh-DioxyBenzP* complexes. Thus, with Rh-*t*-BuBisP* the corresponding hydrogenation products (**85** and **86**) were obtained with 99% ee each, but with different configuration: *R* in case of **85** and *S* in case of **86**.[37,38]

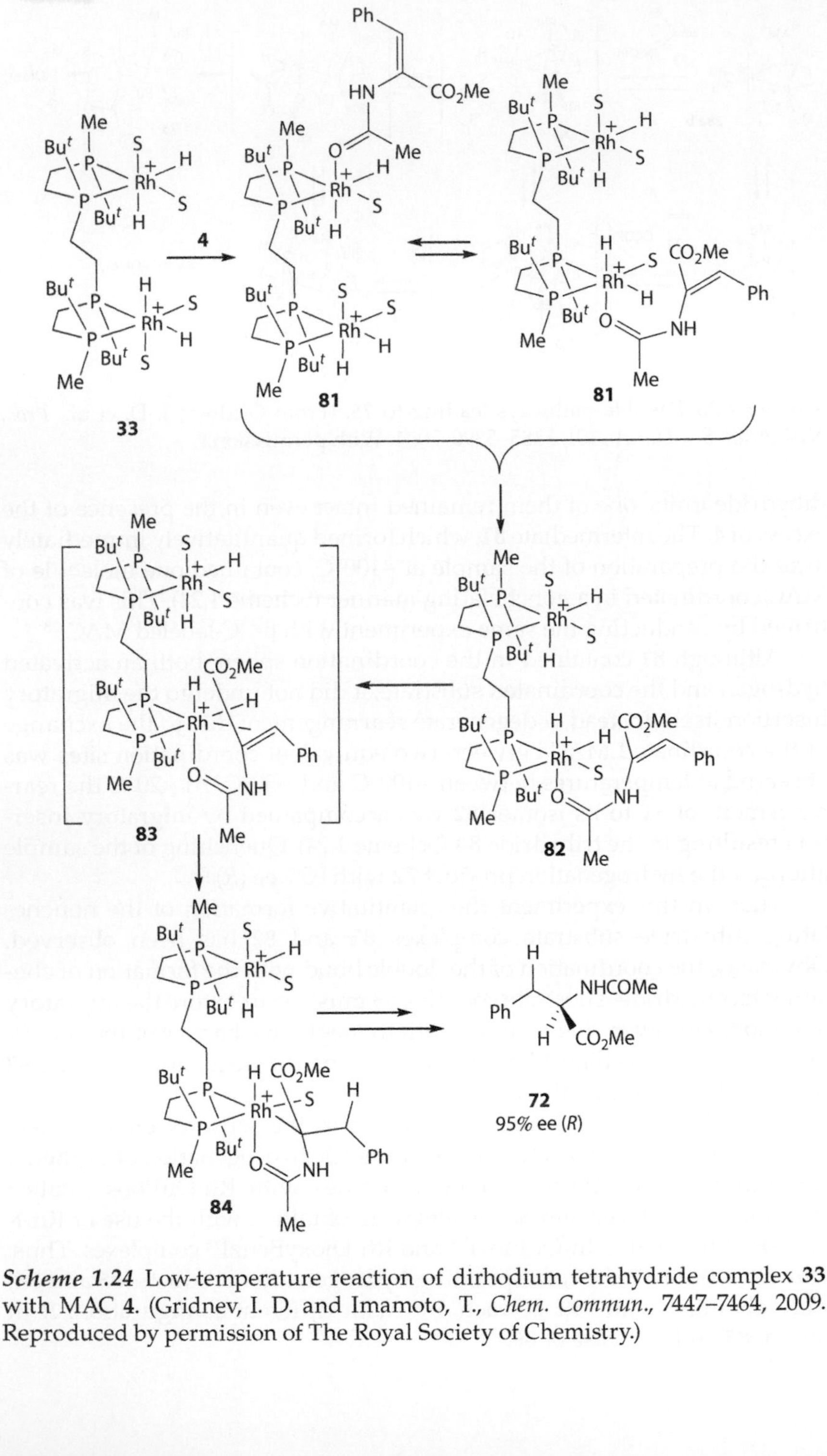

Scheme 1.24 Low-temperature reaction of dirhodium tetrahydride complex **33** with MAC **4**. (Gridnev, I. D. and Imamoto, T., *Chem. Commun.*, 7447–7464, 2009. Reproduced by permission of The Royal Society of Chemistry.)

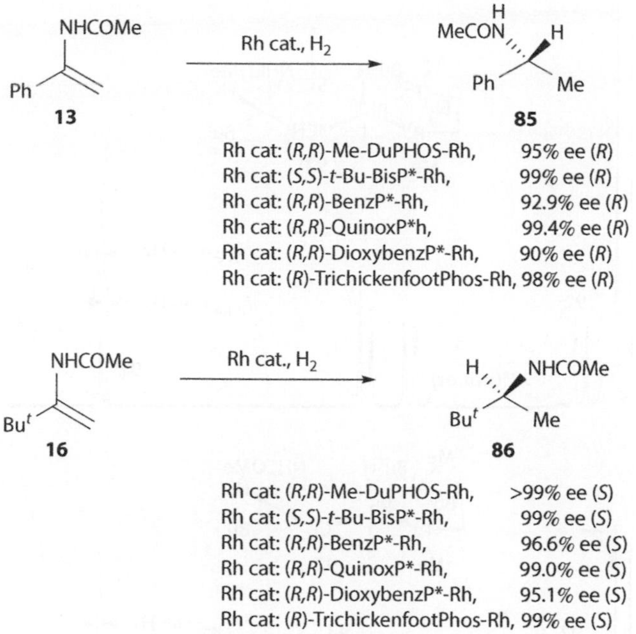
NHCOMe
Ph
13
Rh cat., H2
H
MeCON
H
Ph
Me
85
Rh cat: (R,R)-Me-DuPHOS-Rh, 95% ee (R)
Rh cat: (S,S)-t-Bu-BisP*-Rh, 99% ee (R)
Rh cat: (R,R)-BenzP*-Rh, 92.9% ee (R)
Rh cat: (R,R)-QuinoxP*h, 99.4% ee (R)
Rh cat: (R,R)-DioxybenzP*-Rh, 90% ee (R)
Rh cat: (R)-TrichickenfootPhos-Rh, 98% ee (R)
NHCOMe
Bu^t
16
Rh cat., H2
H
NHCOMe
Bu^t
Me
86
Rh cat: (R,R)-Me-DuPHOS-Rh, >99% ee (S)
Rh cat: (S,S)-t-Bu-BisP*-Rh, 99% ee (S)
Rh cat: (R,R)-BenzP*-Rh, 96.6% ee (S)
Rh cat: (R,R)-QuinoxP*-Rh, 99.0% ee (S)
Rh cat: (R,R)-DioxybenzP*-Rh, 95.1% ee (S)
Rh cat: (R)-TrichickenfootPhos-Rh, 99% ee (S)

Scheme 1.25 Opposite sense of enantioselection observed in the asymmetric hydrogenation of enamides **13** and **16**.

It has been shown by low-temperature NMR experiments, including those using the β-^{13}C labeled substrate, that the opposite sense of enantioselection is explained by the alternative reaction pathways in these two hydrogenations.[37,38] In the case of the bulky enamide **16** it was possible to detect and characterize β-monohydride **92** (Figure 1.5, Scheme 1.26).

In the case of the hydrogenation of **13**, no meaningful intermediates were detected, nevertheless the "normal" reaction pathway was confirmed by analyzing the isotopic distribution in the hydrogenation product, obtained with HD as the reactant. The hydrogenation of **13** (and a series of other aryl-substituted enamides) with HD gave predominantly the product with the deuterium atom in the α-position (with the factor 1.30:1). On the other hand, the HD hydrogenation product of **16** contained more deuterium in the β-position, which correlates well with the original preference of the axial position for deuterium in the solvate dihydride **28** and clearly defines the difference of the reaction pathways (Scheme 1.26).[37,38]

It is clear that, if the complete reversal of the sense of enantioselection was observed, such cases must exist that the competition between the alternative pathways would lead to the low optical yields. Indeed, with *o*-methoxyphenylenamide **94** both types of monohydride intermediates

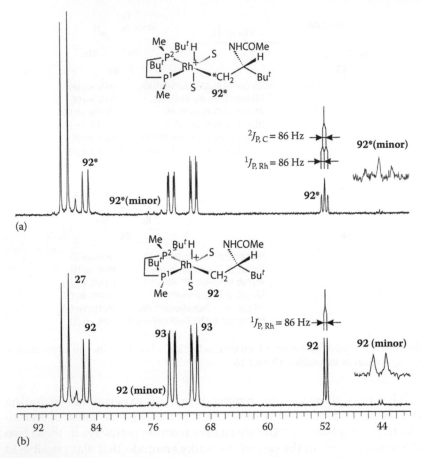

(a)

(b)

Figure 1.5 ^{31}P NMR spectra (162 MHz, −90°C, CD$_3$OD) of the reaction mixtures containing monohydride intermediate **92**: (b) unlabeled: (a) labeled with ^{13}C at the methylene carbon atom. Additional large coupling observed in the case of the labeled compound proves that **92** has a structure of β-monohydride. The minor isomer is evidently also β-monohydride; the ee obtained after quenching of the sample is higher than that calculated based on the relative amounts of major and minor diastereomers. Hence, the major and minor isomers differ in the configuration of the Rh atom. Catalyst–substrate complex **93** was also formed in this experiment. Low-temperature hydrogenation of **93** yields identical spectra. (Reprinted with permission from Gridnev, I. D. et al., *J. Am. Chem. Soc.*, 123, 5268–5276. Copyright 2001 American Chemical Society.)

were detected (Scheme 1.27, Figure 1.6).[38] Comparing the optical yield of the hydrogenation product recovered from the quenched NMR sample with the relative integral intensities of **97–99** in the ^{1}H NMR spectra, it was concluded that **97** and **99** are precursors to the *R*-product, whereas **98** yields **100**(*S*). This result illustrates the competition of three reaction

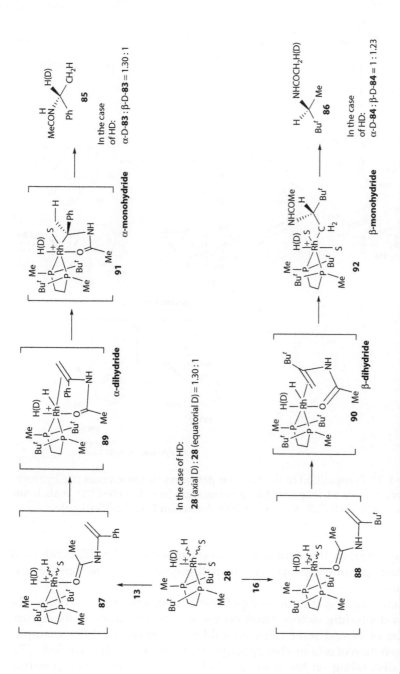

Scheme 1.26 Different reaction pathways in asymmetric hydrogenation of enamides **13** and **16**. (Reprinted with permission from Gridnev, I. D. et al., *J. Am. Chem. Soc.*, 123, 5268–5276. Copyright 2001 American Chemical Society.)

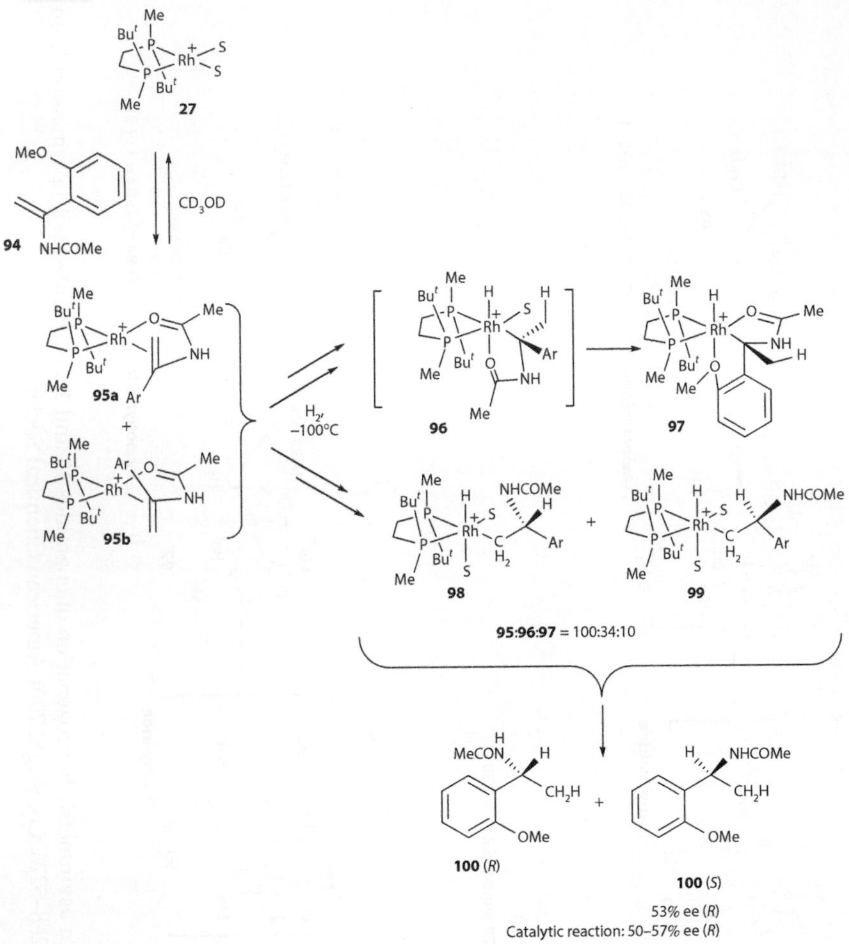

Scheme 1.27 Competition of the reaction pathways in the asymmetric hydrogenation of enamide **94**. (Reprinted with permission from Gridnev, I. D. et al., *J. Am. Chem. Soc.*, 123, 5268–5276. Copyright 2001 American Chemical Society.)

pathways during the double bond coordination in Rh(III) octahedral nonchelating complexes (totally four pathways are available, see the next sections).

Interestingly, due to the competition between the two reaction pathways, a deuterium isotope effect on the enantioselectivity in the hydrogenation of **94** and some other enamides was observed.[44] For example, hydrogenation of α-(*o*-methoxyphenyl)enamide **94** with H_2 gave 50%–57% ee (*R*) depending on the reaction conditions. The same hydrogenation with HD and D_2 afforded the product with 24% ee (*R*) and 5% ee (*S*),

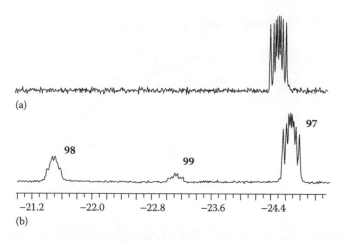

Figure 1.6 Hydride region of the ¹H NMR spectrum (400 MHz, CD₃OD) of the reaction mixture obtained by the low-temperature hydrogenation of the equilibrium mixture of **27, 94, 95a** and **95b**: (b) at –90°C; (a) at –50°C. Assignment of **97** as α-monohydride, and of **98, 99** as β-monohydrides was confirmed by the experiments using ¹³C labeled enamide **51** similar to that described in Figure 1.5. (Reprinted with permission from Gridnev, I. D. et al., *J. Am. Chem. Soc.*, 123, 5268–5276. Copyright 2001 American Chemical Society.)

respectively. This result was explained by considering the hydrogen bonding between Rh–H and the *ortho*-methoxy oxygen atom that makes the pathway through the β-monohydride more favorable.[44]

The β-monohydride intermediates **101a, b** were detected when a low-temperature reaction of **27** with a β-dehydroamino acid, *namely,* methyl (*E*)-3-acetamido-2-butenoate **18** was conducted and in low-temperature hydrogenation of the equilibrium mixture of **3, 18,** and **103** (Scheme 1.28).[40] Similarly to the enamide case, the intermediacy of the β-monohydride correlated with the sense of enantioselection, which was opposite to that observed for α-dehydroamino acids.

1.1.2.6   Catalytic cycle and enantioselective step

Catalytic cycle involving all presently known types of intermediates is shown in Scheme 1.29. The solvate complexes **A**, the catalyst–substrate complexes **C**, and the monohydrides **L** were originally detected at the early stages of the mechanistic studies of Rh-catalyzed asymmetric hydrogenation. Numerous examples of these intermediates were characterized in the following studies. Other intermediates, such as non-chelating catalyst–substrate complexes **B** or solvate molecular hydrogen complexes **D** were detected indirectly through certain effects observed in the NMR spectra. Further characterization of the solvated dihydrides **G**, and occasional detection of such intermediates as the chelating molecular

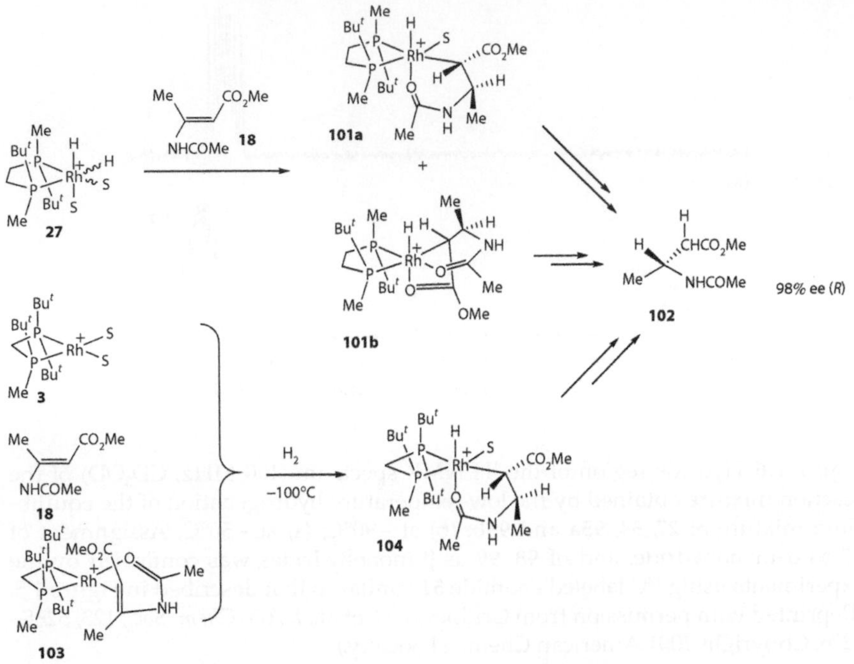

Scheme 1.28 Detection of β-monohydride intermediates in low-temperature reaction of solvate dihydride **27** and ester of β-dehydroamino acid **18**, and in low-temperature hydrogenation of the equilibrium mixture of **3**, **18**, and **103**.

hydrogen complex **77**, the nonchelating octahedral dihydride **81**, and the agostic intermediate **50** makes possible clear understanding that the reaction pool of the Rh-catalyzed asymmetric hydrogenation consists of numerous equilibrating species.

Depending on the nature of the catalyst and the substrate, equilibria can be shifted differently allowing detection of one or another intermediate. Thus, although it is hardly possible to observe all of these intermediates in one catalytic system, there is little doubt that either of them can exist as a low concentrated species under the catalytic conditions.

Furthermore, reversibility of the early steps of the catalytic cycle implies that different pathways are converging in this system. Indeed, the solvate complex **A** can first react either with the substrate yielding **B** or with dihydrogen affording **F**. However, this is not a bifurcation point separating two different mechanisms. An addition of dihydrogen to **B** or of a substrate to molecule **F** would produce the common intermediate **D** that connects numerous interconverting intermediates of the two pathways

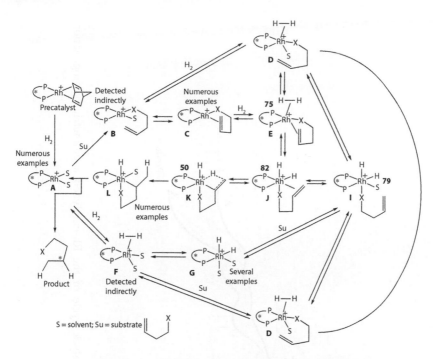

Scheme 1.29 Intermediates involved in the catalytic cycle of asymmetric hydrogenation. (Gridnev, I. D. and Imamoto, T., *Chem. Commun.*, 7447–7464, 2009. Reproduced by permission of The Royal Society of Chemistry.)

into an overall equilibrium. Even the very late intermediate **K** was shown to exist in an equilibrium with the substrate and the solvate complex **A** (*vide supra*).

In view of the above considerations, the handedness of the product must be decided by the mode of coordination of the prochiral double bond taking place in nonchelating Rh(III) octahedral complex **I**. This coordination should be followed by facile migratory insertion and reductive elimination steps, since otherwise the reversibility of all steps before reductive elimination would bring the intermediate **I** back.

The convergence of different pathways is further illustrated in Figure 1.7. Thus, if the catalyst can be hydrogenated itself yielding the solvate dihydride **G** (violet line), then a simple nonchelating coordination of the substrate will produce the key intermediate **I**'. This situation was reproduced in numerous experiments in which the separately prepared solvate dihydrides were reacted with various prochiral substrates.

On the other hand, the oxidative addition to a catalyst substrate complex could be more facile than hydrogenation of the solvate, and

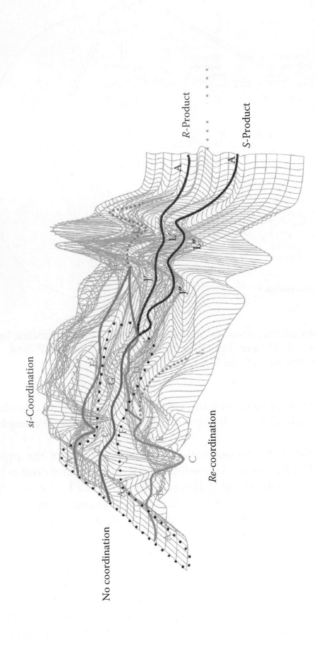

Figure 1.7 Schematic phophile of potential energy for the catalytic cycle of the Rh-catalyzed asymmetric hydrogenation. For explanation see text and Scheme 1.29.

the activation of the dihydrogen could proceed via a catalyst–substrate complex. Taking real examples, a reaction pathway for the hydrogenation of the enamide **13** catalyzed by the Rh–BenzP* complex **8** can be considered (*vide supra*, Scheme 1.4). The oxidative addition to the *re*-coordinated catalyst–substrate complex **C** yields dihydride intermediate **J** via molecular hydrogen complex **E** (light blue line). The migratory insertion in **J** yielding **L** is less facile than the dissociation of the double in **J** that would afford **I**, and for certain pathways isomerization to another diastereomer **I′** might be necessary (*vide infra*).

Similarly, the most facile oxidative addition might involve reaction with dihydrogen with *si-gauche*-coordinated catalyst–substrate complex **C¹** (orange line, *vide supra*, Scheme 1.18). Further transformations leading to the monohydride intermediate **L¹** are relatively facile, but the reductive elimination from the complex **L¹** can require high activation barrier (Scheme 1.19), and instead the reverse reaction yielding **J¹** followed by the double bond dissociation leading to the **I′** would take place.

Hence, in all described cases, the handedness of the product will be determined by the mode of the double bond coordination in the nonchelating octahedral Rh(III) complex **I′** (or its diastereomer, e.g., **I**, *vide infra*) and relative easiness of the following migratory insertion and reductive elimination steps.

1.1.2.7   Process of enantioselection

1.1.2.7.1   Introduction   Totally, there are 12 possible ways of the double bond *cis*-chelating coordination in 2 diastereomers of octahedral nonchelating dihydride intermediates **I** and **I′** in the case of a Rh complex with C_2-symmetric diphosphine ligand. However, already early empirical considerations suggested that the hydride *trans* to phosphorus atom must be transferred in the migratory insertion step.[45] This hypothesis was amply supported by the results of low-temperature experiments (e.g., Figure 1.4), and the structures of all known monohydride intermediates. Hence, the analysis can be restricted to the eight structures originating from two nonchelating complexes **I** and **I′** (Scheme 1.30).

Furthermore, the recent computational data show that the stabilities of the dihydride intermediates **V–VIII** are uniformly lower than these of complexes **I–IV**. Similarly, the energies of the transition states for the following migratory insertion and reductive elimination steps are uniformly higher for the pathways starting from **V to VIII**.[18,20–22,31–34]

Hence, it is possible to restrict the analysis to the four pathways starting from the dihydride intermediates **I to IV**. The complexes **I** and **III** are α-dihydrides yielding opposite enantiomers of the product. Similarly, **II** and **IV** are β-dihydrides yielding the products with opposite handedness.

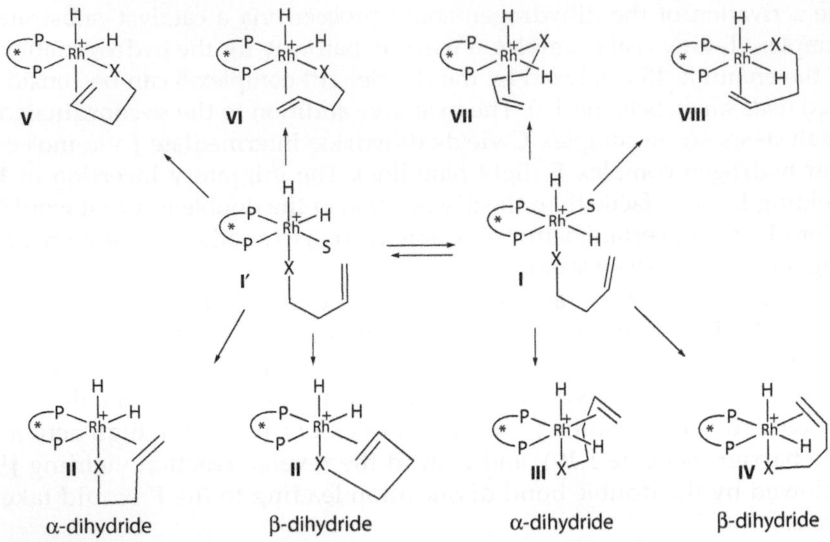

Scheme 1.30 Possible ways for the double bond coordination in I and I'.

1.1.2.7.2 Enantioselection in asymmetric hydrogenation of MAC The first computational analysis of the double bond association in a Rh(III) nonchelating dihydride intermediate was made for the asymmetric hydrogenation of MAC (4) with Rh-(R)-trichickenfootPhos 3.[12] Only α-dihydride pathways proceeding via the formation of the chelate cycle nearby the chiral phosphorus atom have been computed. Actual displacement of a methanol molecule from the Rh coordination site was modeled via coordinating it with the carboxymethyl group of the substrate (Scheme 1.31, Figure 1.8).[12]

The transformation of **105a–106a** implies changing the lobe of the oxygen atom used for making the coordination bond with rhodium with simultaneous pushing the molecule of methanol out of the coordination sphere through the **TS9a**. This process requires a lot of activity in the less hindered quadrant, and the MeOH molecule can easily get out of the way via making a hydrogen bond with amide NH moiety (Figure 1.8).

A same process is impossible for the transformation of **105b–106b** due to the hindrance provided by the *t*-butyl group, and the coordination of the double bond must take very different pathway accompanied by pushing the methanol molecule toward the less hindered quadrant (Figure 1.9).

Similar analysis was made for the catalytic cycle of asymmetric hydrogenation of 4 with the Rh-TangPhos catalyst (Figure 1.10).[22] In the case of formation of the chelate cycle in the less hindered quadrant, the

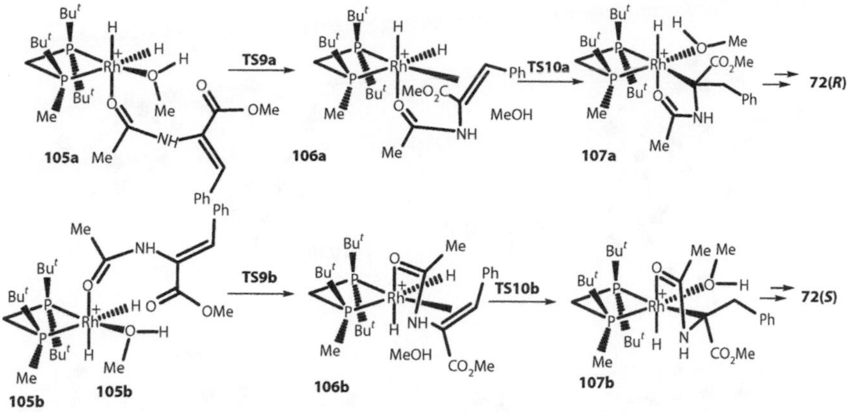

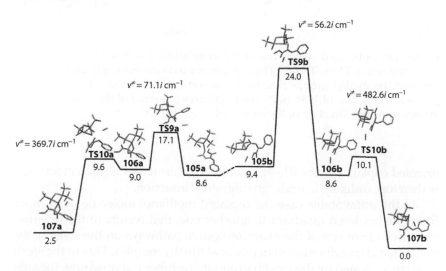

Scheme 1.31 Double bond coordination and migratory insertion steps in the nonchelating dihydride complexes **105a** and **105b**. (Reprinted with permission from Gridnev, I. et al., *J. Am. Chem. Soc.*, 130, 2560–2572. Copyright 2008 American Chemical Society.)

Figure 1.8 Pathways for the double bond coordination and migratory insertion steps in the nonchelating complexes **105a** and **105b** computed on the B3LYP/SDD level of theory. (Reprinted with permission from Gridnev, I. et al., *J. Am. Chem. Soc.*, 130, 2560–2572. Copyright 2008 American Chemical Society.)

scan proceeds via a single maximum at approximately 3 Å and results in the corresponding dihydride intermediate. On the other hand, in the case when the chelate cycle is formed in the hindered quadrant, the maximum reached only at 2.5 Å has unreasonably high energy. The minimum reached upon further approach of the double bond has the double bond

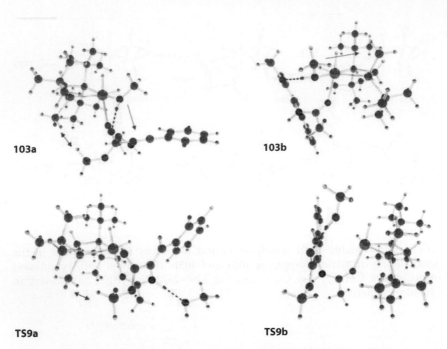

103a   103b

TS9a   TS9b

Figure 1.9 Optimized structures of the nonchelating dihydrides **105a, b** and transition states **TS9a, TS9b**.[12] The red arrows indicate relatively close contacts between the methyl groups that would make impossible the same association pathway in the case of **105b** (with the P–Bu^t group instead of the P–Me); the blue arrows show the direction of the methanol displacement.

oriented coplanar to the Rh–H axial bond (*trans* to the oxygen atom), and is therefore unlikely to undergo migratory insertion.

In the unfavorable case the replaced methanol molecule must move from one hindered quadrant to another one, that results in the by effective blockade of one of the enantioselective pathways on the stage of the double bond coordination in octahedral Rh(III) complex. Due to the steric hindrance created by the *tert*-Bu group in the hindered quadrant, the substitution of the solvent molecule by the double bond becomes practically impossible.

Accurate conclusions on the sense and level of enantioselection can be made also by analyzing the double bond coordination step without explicit computation of the solvent molecules (e.g., Figure 1.11). In this case, application of the latest available functionals accurately describing dispersion allows us to make some interesting conclusions on the interplay between attractive and repulsive interactions between substituents during the enantioselective step.

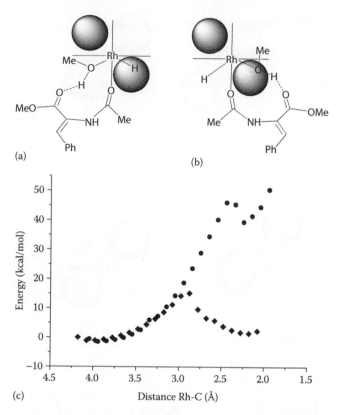

(a)

(b)

(c)

Figure 1.10 Scan of relative energies of the double bond replacing the MeOH molecule at the Rh coordination site: chelate cycle formed in a less hindered quadrant (squares); chelate cycle formed in a hindered quadrant (circles). Computed at the B3LYP/SDD(Rh)/6-31G(d, p)(all others)/CPCM(MeOH) level of theory with additional diffuse function for phosphorus. (Gridnev, I. D. et al., *Dalton Trans.*, 43, 1785–1790, 2014. Reproduced by permission of The Royal Society of Chemistry.)

When the distance between the Rh atom and the quaternary carbon atom of the double bond is 4.5 Å, both diastereomeric nonchelating complexes have the same stability. Analysis of the structures with this Rh–C distance shows that there are evident nonclassical attractive interactions between the substituents on the phosphorus atom and carboxymethyl group in one case (C–H...O interactions) or phenyl group in another case (C–H...π interactions).

In the first case, the approach of the double bond to the Rh atom is accompanied by the simultaneous approach of the carboxymethyl group to the alkyl substituents on the phosphorus atom, and strengthening of the C–H...O interactions. This results in achieving the minimum at 4.0 Å, and then two maxima at 3.5 Å and 2.7 Å that occur because of

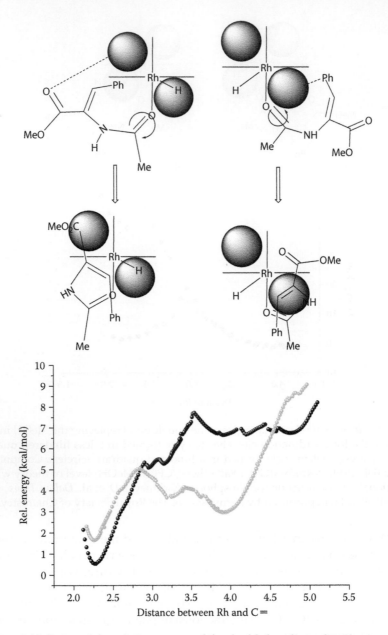

Figure 1.11 Scans of the relative energy of the double bond coordination to a free
coordination site of the nonchelating Rh(III) octahedral complexes. Computations
were done on the B3LYP-D3/6-31G*/SMD(methanol) level of theory. Gray spheres:
coordination drawn in the upper left Scheme; black spheres: coordination drawn
in the upper right Scheme.

switching of the strongest C–H...O interactions from the methyl group to the *tert*-Bu group, and from one oxygen atom of the carboxymethyl group to another one.

On the other hand, when the chelate cycle is formed in the "hindered" quadrant, the strong C–H...π interactions keep the double bond in a conformation preventing effective stabilization of the molecule via its interaction with the Rh atom even though the distance between Rh and the quaternary carbon atom is reducing from 4.5 to 3.5 Å. Only after these C–H...π interactions become impossible due to the increasing distance between the phenyl group and the substituents on phosphorus (maximum at 3.5 Å), the further approach of the quaternary carbon atom to the Rh atom results in increasing stability of the species.

One way to make a quantitative estimation of the optical yield in this case is comparing the energy of the maxima at 2.7 Å in the first case and 3.5 Å in the second case that gives a reasonable value of ΔE around 3 kcal/mol. On the other hand, one can also conclude that the extremely effective stereoselection can take place earlier, because the relative concentration of the diastereomeric species with the Rh–C distance of 4.0 Å would be determined by a significantly larger ΔE (about 4 kcal/mol).

In other words, the "attractive" C–H...π interactions between the *tert*-Bu group of the catalyst and the phenyl group of the substrate effectively "hinders" approach of the double bond to the Rh atom.

Noteworthy, this mechanism of enantioselection explains well the absence of the temperature effect on the optical yield in the asymmetric hydrogenation of **4** catalyzed by the Rh-TangPhos catalyst. Indeed, in the case when the ee is determined by the difference in the free energies of the diastereomeric transition states, significant dependence of the optical yields on the temperature is usually observed due to the natural temperature variation of the rate constants. On the other hand, the structure of the catalyst implying facile formation of the chelate cycle in a less hindered quadrant and impossibility of similar transformation with the chelate cycle forming in hindered quadrant, remains the same at any temperature. Hence, the almost perfect enantioselectivity can be observed either in catalytic or low-temperature stoichiometric reaction (>99% ee in both cases).

1.1.2.7.3 Enantioselection in asymmetric hydrogenation of esters of β-dehydroamino acids Computations of the enantioselective step in the asymmetric hydrogenation of five representative esters of β-dehydroamino acids were carried out recently. Competition of two β-dihydride pathways and two unsaturated pathways was considered (Scheme 1.32, Figure 1.12). In Figure 1.12 one can see that there are three characteristic values that determine the enantioselectivity of the catalytic reaction: ΔG^1, difference in

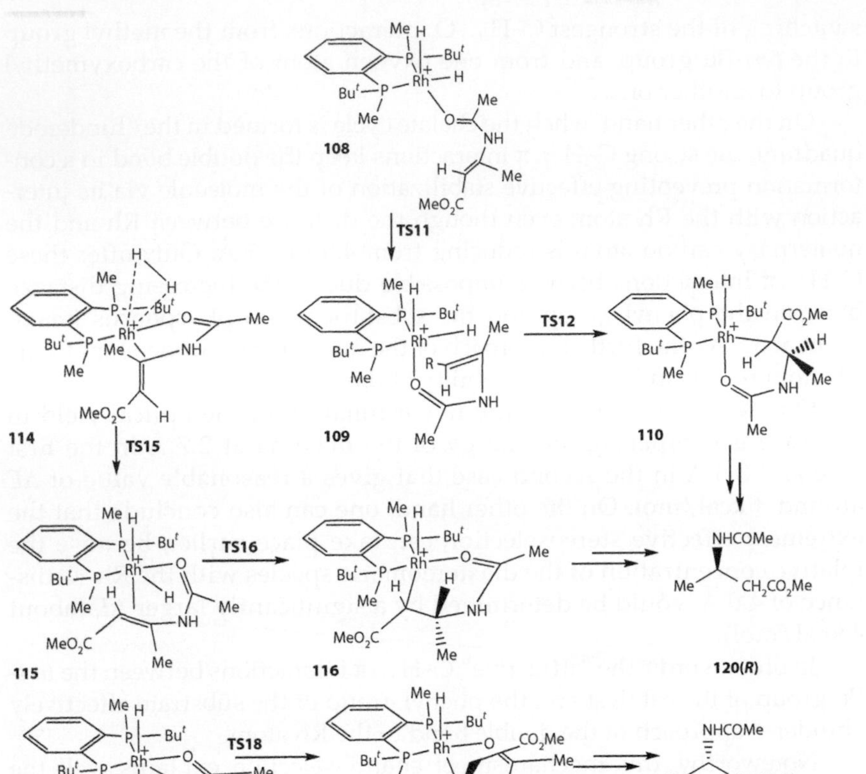

Scheme 1.32 Four computed pathways for the formation of the products **120(R)** and **120(S)** in hydrogenation of **26** catalyzed by Rh-BenzP* complex **8**. (Reprinted with permission from Gridnev, I. D. et al., *ACS Catal.*, 4, 203–219. Copyright 2014 American Chemical Society.)

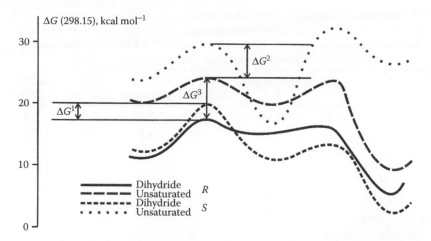

Figure 1.12 Sections of profiles of free energy for the enantioselective stages of the four different catalytic pathways in hydrogenation of **26** catalyzed by **8**. (Reprinted with permission from Gridnev, I. D. et al., *ACS Catal.*, 4, 203–219. Copyright 2014 American Chemical Society.)

the stabilities of the transition states for the double bond coordination **TS11** and **TS13**; ΔG^2, difference in the stabilities of the transition states for the oxidative addition of H_2 **TS15** and **TS17**; ΔG^3, difference in the stabilities of the stereodetermining transition states for dihydride and unsaturated mechanisms.

The computed values of ΔG^1, ΔG^2, and ΔG^3 for the hydrogenation of five different substrates catalyzed by BenzP*–Rh complex **8** are collected in Table 1.2. The expected values ΔG^{exp} were derived from the optical yields obtained experimentally in catalytic hydrogenation at ambient temperature and in low-temperature hydrogenations (Table 1.2).

In the case of the hydrogenation of **26** the transition state for the oxidative addition of H_2 in the unsaturated pathway was computed to be 7.2 kcal/mol less stable than the transition state of the double bond coordination. This value only slightly decreases at 173 K, hence the unsaturated pathway is not interfering with the dihydride route in this case at any temperature. Therefore, the optical yield of the catalytic hydrogenation of **2a** catalyzed by **5** is determined solely by the value of ΔG^1, which was computed to be virtually the same at 298 K and at 173 K. Although the ΔG^{exp} in this case is notably higher than computed ΔG^1, the computations reliably reproduce R-enantioselective reaction with the optical yield of about 96% ee which is not affected by the temperature changes (see also below).

Table 1.2 Computed values of the free energy differences between corresponding transition states in the catalytic pathways of hydrogenation of various substrates catalyzed by BenzP*–Rh complex **8** and expected values derived from the experimental ee values (see text for details)

Substrate	Temperature, K	ΔG^1, kcal/mol	ΔG^2, kcal/mol	ΔG^3, kcal/mol	ΔG^{exp}, kcal/mol
26	298.15	2.7	5.0	7.2	3.8
	173.15	2.6	4.8	5.6	3.8
18	298.15	1.9	−0.6	3.7	2.6
	173.15	2.2	−0.6	2.9	1.1
19	298.15	2.8	0.9	4.9	2.5
	173.15	2.5	0.8	3.9	2.3
20	298.15	1.5	−1.6	6.5	2.3
	173.15	1.8	−1.6	5.4	2.5
21	298.15	1.5	0.7	4.0	1.1
	173.15	1.9	0.3	3.1	1.7

Source: Gridnev, I. D. et al., *ACS Catal.*, 4, 203–219, 2014.

On the other hand, the computed ΔG^3, free energy gap between dihydride and unsaturated mechanisms, is notably smaller for the hydrogenation of **18** (Table 1.2). At 173 K it is less than 3 kcal/mol, and the interference of the nonstereoselective unsaturated pathway cannot be neglected. Since, the unsaturated pathway is slightly *S*-stereoselective in this case, these computational results may explain the dramatic decrease of the ee observed in the low-temperature hydrogenation of **18** compared to the catalytic reaction at ambient temperature.

The experimental data are nicely reproduced computationally for substrate **19**. Although the unsaturated mechanism in this case is just moderately *R*-stereoselective, it probably only marginally interferes with the dihydride mechanism either at 298 or 173 K and the order of enantioselection is mostly determined by the ΔG^1 values, which are quite close to the corresponding ΔG^{exp}.

The structures of the rate- and stereodetermining transition states **TS19** and **TS20** are compared in Figure 1.13 (transition states for the double bond coordination in asymmetric hydrogenation of **19** catalyzed by **8**). Both transition states are characterized by relatively low absolute

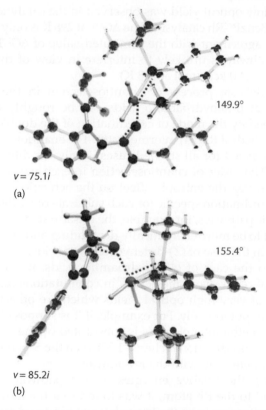

149.9°

$v = 75.1i$

(a)

155.4°

$v = 85.2i$

(b)

Figure 1.13 Optimized structures of **TS19** (a, leads to the formation of *R*-product) and **TS20** (b, leads to formation of *S*-product). The larger dihedral angle is observed in **TS20**, because the forming chelating cycle avoids close approach to the Bu^t group. (Reprinted with permission from Gridnev, I. D. et al., *ACS Catal.*, 4, 203–219. Copyright 2014 American Chemical Society.)

values of the imaginary frequencies normal for the intramolecular movement. In both cases the largest displacement vectors correspond to the approach of the CH=C unit to the H–Rh–H plane. In the case of the **TS19** the formation of the chelate cycle occurs in the less hindered quadrant. In the case of **TS20** the Bu^t substituent hinders the formation of the chelate cycle that results in a transition state destabilized for 2.8 kcal/mol.

The computed values of ΔG^1 for the hydrogenation of **20** (1.5 and 1.8 kcal/mol at 298 K and 173 K, respectively) are somewhat lower than the corresponding ΔG^{exp} which are 2.3 and 2.5 kcal/mol. Nevertheless, the enantioselectivity is correctly predicted. The unsaturated pathway is notably *S*-stereoselective in this case, but is not competing with the dihydride pathway, since ΔG^3 values are quite high (6.5 kcal/mol at 298 K and 5.4 kcal/mol at 173 K).

Relatively low optical yield was observed in the catalytic hydrogenation of **21** with BenzP*-Rh catalyst **8**. The ΔG^{exp} at 298 K is only 1.1 kcal/mol in a reasonable agreement with the computed value of ΔG^1 1.5 kcal/mol. Unsaturated pathway is unlikely to interfere in view of the reasonably high value of ΔG^3 (4.0 kcal/mol at 298 K).

Thus, the delicate process of enantioselection in the asymmetric hydrogenation of β-dehydroamino acids can be roughly described as being determined by the mode of coordination of the double bond of the substrate in octahedral Rh(III) intermediates. The enantioselectivity could be correctly predicted for all studied cases if only the dihyride pathway is considered. The order of enantioselection in each particular case can be affected either by the entropic effect on the activation barrier for the double bond coordination specific for each substrate or by the interference of other catalytic pathways, for example, the unsaturated route, which has been computed to be more energetically demanding and significantly less stereoselective in the case of (Z)-β-dehydroamino acids.

Similarly to the case of α-dehydroamino acids, it was shown that incorporation of real solvent molecules in computations makes possible rationalization of very high optical yields which are otherwise difficult to reproduce computationally. For example, if it is supposed that a linear cluster of three methanol molecules is substituted by the double bond in the process of its coordination (Scheme 1.33), then the *S*-pathway is hardly expected to compete at all with the *R*-pathway.

By scanning the relative energies on the pathways approaching the double bond to the Rh atom, it was found that the movement of the methanol cluster occurs from the lower left less hindered quadrant to the upper right less hindered quadrant in the transformation of **121** to **122**. On the contrary, the conversion of **123–124** requires continuous movement of the methanol cluster within the hindered quadrants. Accordingly, the energy scans look out quite differently (Figure 1.13). The scan of the

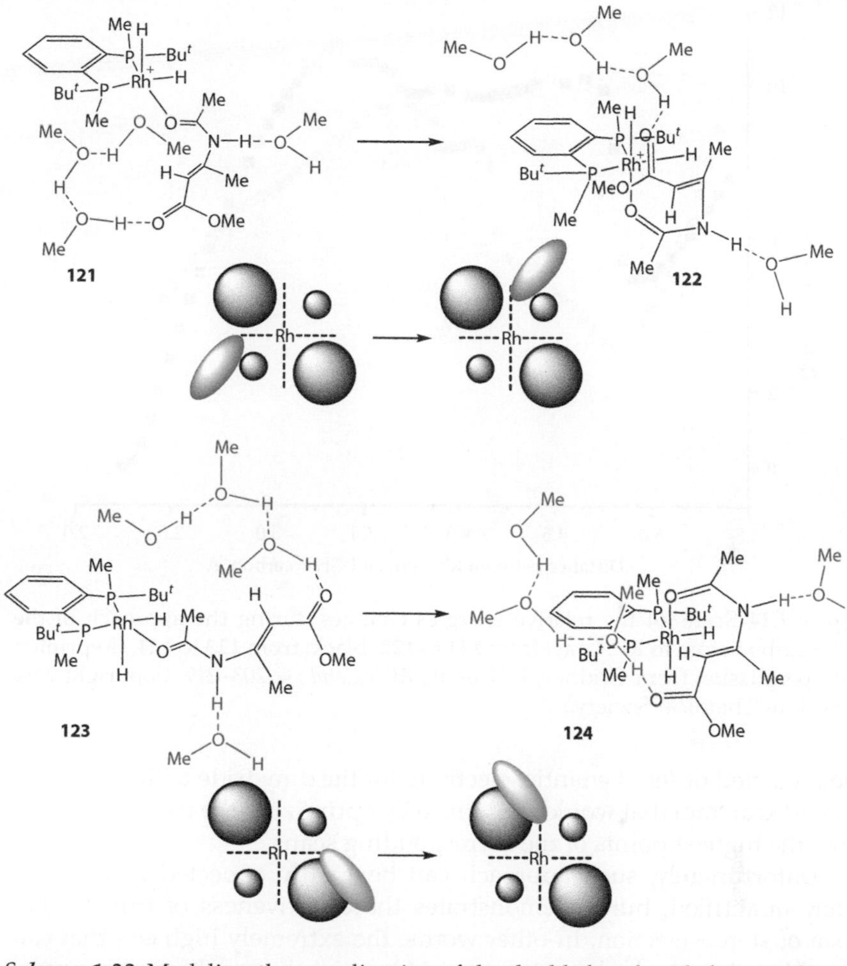

Scheme 1.33 Modeling the coordination of the double bond with four methanol molecules. (Reprinted with permission from Gridnev, I. D. et al., *ACS Catal.*, 4, 203–219. Copyright 2014 American Chemical Society.)

former conversion can be roughly described as a potential energy profile of a single elementary step, whereas in the latter case the energy raises sharply when the C–Rh distance reaches about 5 Å and remains constantly high until 3 Å.

Taking into account that all previous steps are reversible, most of the molecules with CH= atom being 4.5 Å far away from Rh or closer will have conformation ultimately leading to **122**. More precisely, the relative concentrations of such molecules preceding **122** and preceding **124** will be determined by Boltzmann distribution for the energy difference of approximately 5 kcal/mol (Figure 1.14). Hence, this value will specify

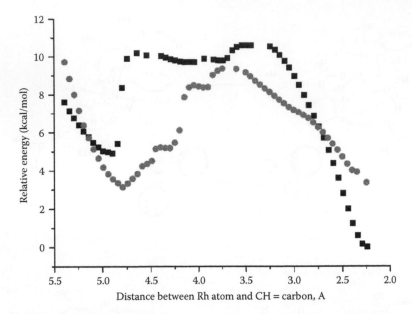

Figure 1.14 Scans of the relative energies changes during the approach of the CH = carbon atom to Rh. Gray: from **121** to **122**. Black: from **123** to **124**. (Reprinted with permission from Gridnev, I. D. et al., *ACS Catal.*, 4, 203–219. Copyright 2014 American Chemical Society.)

the predicted order of enantioselectivity for the dihydride pathway, rather than ~2 kcal/mol that would be derived by optimizing the transition states from the highest points of the corresponding scans.

Unfortunately, such approach can be hardly expected to be accurately quantified, but it demonstrates the effectiveness of this mechanism of stereoselection. In other words, the extremely high ee's that can be achieved in the hydrogenation of **26** are not the result of a "fair competition" between the rates of similar elementary stages, but the result of the effectively blocked approach to this elementary stage leading to one of the enantiomers. Remarkably, the effectiveness of this approach is not expected to depend significantly on the temperature. Hence, the uniformly high ee's in catalytic hydrogenations and low-temperature experiments (as in the case of **26** hydrogenated with **3** or **8**) can be considered as the evidence of purely dihydride mechanism. Otherwise, if significant changes of the optical yields with the temperature are observed, this testifies for the existence of competing mechanisms.

1.1.2.7.4   Enantioselection in asymmetric hydrogenation of enamides   The detailed computations of the catalytic cycles for hydrogenation of five different substrates catalyzed by Rh–BenzP* complex convincingly showed

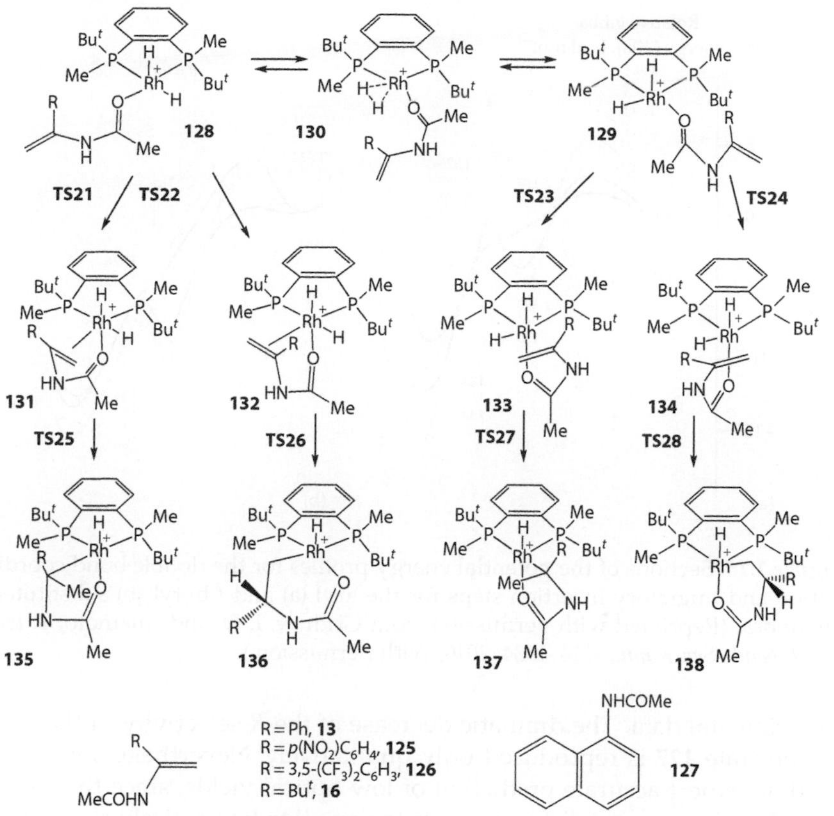

Scheme 1.34 Enantioselective stages in the asymmetric hydrogenation of enamides. (From Gridnev, I. D. and Imamoto, T. *Izv. Akad. Nauk, Ser. Khim.*, 1514–1534, 2016. With permission.)

that enantioselection takes place via the combination of the double bond coordination and migratory insertion stages (Scheme 1.34).[21]

In the case of the aryl-substituted enamides **13, 125–127** (see Figure 1.15a, exemplified for substrate **13**), the most facile mode of the double bond coordination proceeds via the **TS21** and leads to the formation of α-dihydride **131** with the chelating cycle positioned in a less hindered quadrant. It competes with the formation via the **TS22** of a β-dihydride **132** which also has its chelating cycle positioned in the less hindered quadrant. Hence, the level of the enantioselection must be determined by the difference ΔG^1 in stabilities of **TS21** and **TS22** (e.g., Figure 1.15a).

Table 1.3 shows that the computational analysis provided a very good qualitative prediction of the sense and level of enantioselection. The high *R*-selectivity is predicted for the substrates **13, 125, 126** in accord with the

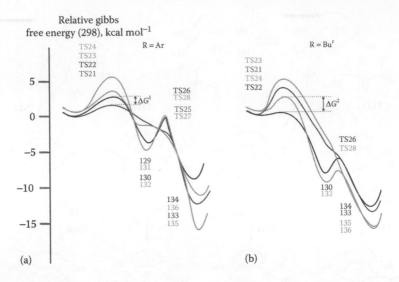

Figure 1.15 Sections of the potential energy profiles for the double bond coordination and migratory insertion steps for the aryl (a) and *t*-butyl (b) substituted enamides. (Reprinted with permission from Gridnev, I. D. and Imamoto, T. *Izv. Akad. Nauk, Ser. Khim.*, 1514–1534, 2016. With permission.)

experimental data. The dramatic decrease of the *R*-selectivity in the case of substrate **127** is reproduced only qualitatively. Nevertheless, it is difficult to expect accurate prediction of low optical yields, since the corresponding free energy differences are too small to be precisely computed. On the other hand, the accurate prediction of the complete enantioreversal in the case of the *t*-butyl-substituted enamide **16** is very gratifying. In that case two pathways proceeding through β-dihydrides **132** and **134** are competing, and the enantioselectivity is determined by the value of ΔG^2 (Figure 1.15b).

Figure 1.16 shows the optimized structures of the transition states of two competing pathways for the double bond coordination. Since both TS's lead to the formation of the chelate cycle in the less hindered quadrant, the relative stability of **TS21** is evidently stipulated by C–H...π interaction between the substituted phenyl ring of the substrate and the *t*-Bu substituent of the catalyst, similar to that experimentally detected in the corresponding square planar catalyst–substrate complex. This conclusion is in accord with the experimentally observed increase of the optical yield of the asymmetric hydrogenation in a series **13**<**123**<**124**, although we were unable to reproduce this effect computationally.

In the case of the *t*-butyl-substituted enamide **16**, the α-dihydride pathway is disfavored due to the impossibility of the C–H...π interaction and significant steric hindrance between the two adjacent *t*-Bu groups,

Table 1.3 Comparison of the experimental and computed optical yields (ee, %) in the Rh-catalyzed asymmetric hydrogenation of enamides **13**, **16**, and **123–125**

Substrate	13	125	126	16	127
ΔG^1, kcal/mol	1.7	1.9	1.7	–	0.2
ΔG^2, kcal/mol	–	–	–	2.3	–
Computed ee	89.4 (R)	91.6 (R)	89.4 (R)	94.6 (S)	16 (R)
Experimental ee, 298 K	89.6 (R)	99.2 (R)	99.9 (R)	96.6 (S)	53 (R)
Experimental ee, 193 K	98.9 (R)	99.2 (R)	98.6 (R)	–	–

Source: Gridnev, I. D. and Imamoto, T. *Izv. Akad. Nauk, Ser. Khim.*, 1514–1534, 2016.

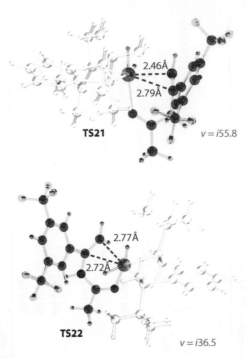

Figure 1.16 Optimized structures of the **TS21** and **TS22** in the case of substrate **13**. (From Gridnev, I. D. and Imamoto, T. *Izv. Akad. Nauk, Ser. Khim.,* 1514–1534, 2016. With permission.)

and the level of the stereochemical induction is determined via competition of the two β-dihydride pathways (Figure 1.17). Either of the TS's shown in Figure 1.17 does not exhibit any steric hindrance, and hence the computed Gibbs free energy difference of 2.3 kcal/mol can be considered as an evaluation of the relative stability of the β-chelate ring formation in the less hindered quadrant of the catalyst. Apparently, the **TS28** is a much "earlier" TS than **TS26,** in which the stereoselection can occur at the initial period of the approaching of the double bond to the Rh atom.

1.1.2.8   Sense of enantioselection and the relative size of substituents

In the previous sections, it has been shown that the intuitive ideas on the relative size of the substituents of the substrate are rarely helpful for the estimation of the relative stabilities of intermediates or transition states even when these substituents are actually different in shape and size.

This is still more difficult to do in the case of the diphosphine ligands with backbone chirality, when all four substituents on the phosphorus atoms are phenyls. The conformation of the chelate cycle (fixed by the backbone substituents) makes these four phenyls pairwise nonequivalent

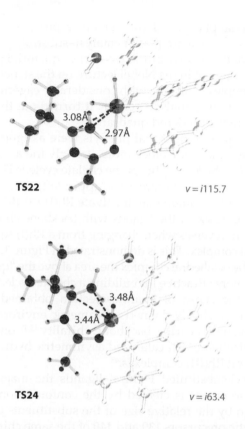

Figure 1.17 Optimized structures of the **TS22** and **TS24** in the case of substrate **16**. (From Gridnev, I. D. and Imamoto, T. *Izv. Akad. Nauk, Ser. Khim.*, 1514–1534, 2016. With permission.)

("pseudo"-axial and "pseudo"-equatorial), thus securing the C_2-symmetrical environment around the rhodium atom essential to accomplish the asymmetric catalysis. However, the question of which of these two types of spatially nonequivalent phenyls provides the necessary hindrance to realize stereoselection was a source of a considerable controversy.

Originally, the question emerged when Knowles introduced his famous quadrant diagrams as an illustration of the empirical correlation between the spatial arrangement of the phenyls in the known X-ray structures of the various catalytic precursors.[46] The correlation works well in predicting the chirality of the product based on the chirality of the catalytic precursor.[40] The early mechanistic interpretation of the quadrant rule in terms of the effective size of the substituents was inspired by the spectacular leaning of the equatorial phenyls from the plane orthogonal to the chelate cycle of the known catalytic precursors, and the fact that these phenyls are "edge"-oriented in their solid state, contrary to the "face"-oriented axial phenyls.[47] However, an empirical explanation of the relative stability of the square planar catalyst–substrate complexes based on the proposal of the relative bulkiness of the equatorial phenyls did not work. As Knowles puts it in his Nobel lecture on the structure of the catalyst–substrate complex, "…it was with considerable eagerness we awaited the X-ray crystallograpic analysis results. It turned out that the enamide was lying nicely in the hindered quadrant."[48]

Initially the "quasi"-equatorial phenyls were assigned as effectively large substituents, since they block more strongly the area in front of the plane orthogonal to the Rh–diphosphine chelate cycle.[46] This stereochemical feature is illustrated in Figure 1.18 (upper left) that shows the optimized structure for the square planar solvate Rh(I)-(S)-BINAP complex.

However, in the case of the ligands with backbone chirality the asymmetric environment reverses when changing from a Rh(I) square planar to a Rh(III) octahedral complex. This is demonstrated in Figure 1.18 (bottom left), showing that in the octahedral complex the area above the "quasi"-equatorial phenyl becomes more attractive for building a chelate cycle.[43]

Comparing the sense of enantioselection obtained with P-chiral ligands generating a clearly defined asymmetric environment with that observed with ligands bearing backbone chirality,[42,49] one can conclude that the stereoselection in Rh-catalyzed asymmetric hydrogenation takes place in octahedral Rh(III) complexes.[43]

With tetraaryl-substituted P-chiral ligands the major effect on the asymmetric environment is created by the conformation of the chelate cycle, rather than by the relative size of the substituents on phosphorus. Thus, two catalytic precursors **139** and **140** of the same chiral diphosphine can have two opposite arrangements of the phenyl and *o*-tolyl rings in the solid state, and still (quite unsurprisingly) yield the hydrogenation product with the same sense of enantioselection (Figure 1.19).[50]

This means that both **139** and **140** give the same active catalyst species **141**. That is, the backbone conformation is changed from that observed in the solid state. The *ortho*-tolyl substituents are larger than phenyls, and in solution they occupy "quasi"-equatorial positions to reduce the 1,3-interactions between the aryl group and the methylene hydrogen atom. This makes the phenyls that are actually smaller in size to create the hindrance for the chelation in the area above and below of the Rh-diphosphine chelate cycle.

The latter conclusion helps to understand the sense of enantioselection obtained in asymmetric hydrogenations catalyzed by Rh complexes of DIPAMP and structurally similar ligands. The origin of the high stereogenic potential of the DIPAMP ligand has long fascinated researchers in this field. Indeed, the difference in size between *o*-anisyl and phenyl groups seems too insignificant for the excellent stereoselection observed in the hydrogenations catalyzed by Rh–DIPAMP complex. Knowles

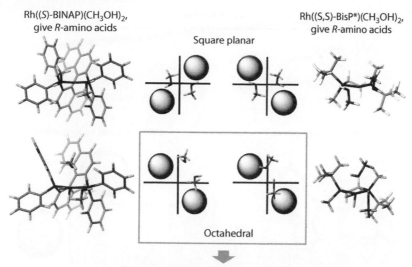

Figure 1.18 Optimized structures (B3LYP/SDD) of the square planar solvate Rh(I) complexes [Rh((S)-BINAP)(CH$_3$OH)$_2$]$^+$ and [Rh((S, S)-t-Bu-BisP*)(CH$_3$OH)$_2$]$^+$ (top), and octahedral solvate dihydride Rh(III) complexes [RhH$_2$((S)-BINAP) (CH$_3$OH)$_2$]$^+$ and [RhH$_2$((S, S)-t-Bu-BisP*)(CH$_3$OH)$_2$]$^+$ (below). The orientation of the methanol molecules helps to determine the less hindered quadrant in each case (in octahedral complexes the equatorially coordinated methanol molecule is brought away from the substituents on the phosphorus atoms, hence only the position of the axially coordinated methanol molecule is informative). In case of the P-chiral t-Bu-BisP* complexes (right), the quadrant diagram is the same for Rh(I) and Rh(III) because the relative size of the substituents always remains the same. Hence, it can be concluded that when a large substituent is in the upper left corner, R-amino acids are produced. In the case of tetraphenyl-substituted BINAP complexes, the quadrant diagram is opposite for square planar and octahedral complexes. The Rh-(S)-BINAP complex gives R-amino acids in asymmetric hydrogenation, hence the arrangement of the larger and smaller substituents in the stereodiscriminating step must be the same as in the t-Bu-BisP* case. Thus, one can conclude from these data that stereoselection takes place in octahedral complexes. (Gridnev, I. D. and Imamoto, T., *Chem. Commun.*, 7447–7464, 2009. Reproduced by permission of The Royal Society of Chemistry.)

himself suggested that the *ortho*-methoxy group of the anisyl substituent might participate in the coordination of the DIPAMP ligand to rhodium, thus securing the higher stereoselection.[51]

However, analysis of the hydrogenation results for the Rh complexes of DIPAMP (**142**) and a series of DIPAMP-like diphosphine ligands **143–146** (Table 1.4) shows that the very similar ee's are achieved with the ligands that do not have any donor atoms capable for additional coordination in

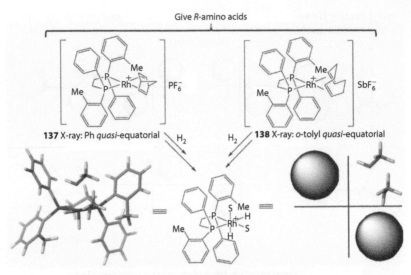

139 in stable conformation, *o*-tolyl *quasi*-equatorial

Figure 1.19 In the case of P-chiral ligands with flat aryl substituents, the relative size of the substituents is determined conformationally. (Gridnev, I. D. and Imamoto, T., *Chem. Commun.*, 7447–7464, 2009. Reproduced by permission of The Royal Society of Chemistry.)

the rhodium complexes.[50] Moreover, the results shown in the Table 1.4 demonstrate the uniformity of the sense of enantioselection observed in asymmetric hydrogenations utilizing these ligands; if the "larger" substituted phenyl ring is regarded as a bulky substituent, then predictions of the quadrant rule are opposite for the P-stereogenic ligands with clear difference in the size of the substituents on phosphorus.[42]

The conformation of the chelate cycle of the catalyst in solution must be determined in the Rh complexes of the ligands **142–146** by the substituents on phosphorus. The substituted phenyls would therefore preferentially occupy the equatorial positions. This is confirmed by the X-ray structure for the rhodium complex of **146** (Figure 1.20).[42] In solution, the tetrahydronaphthalene substituents can easily acquire a conformation where they would not create any hindrance above the chelate cycle, whereas the axial positions of the phenyls would be fixed by the conformational locks (Figure 1.20b). Accepting this line of argument, it can be concluded that the unsubstituted phenyls in the Rh complex of **146** act as stereoregulating substituents. This, in turn, gives a quadrant diagram that coincides with the general quadrant rule (see above). In other words, being formally P-stereogenic ligands, DIPAMP (**142**) and its analogs (**143–146**) work as the catalysts with backbone chirality in asymmetric hydrogenation (Figure 1.20).

Table 1.4 Results of asymmetric hydrogenations (ee [configuration of the product]) catalyzed by the Rh complexes of diphosphines **142–146**

Ligand	Ph–CH=C(NHCOMe)CO₂Me	Ph–CH=C(NHCOMe)CO₂H	AcO/MeO-aryl–CH=C(NHCOMe)CO₂H	CH₂=C(NHCOMe)CO₂H	Refs.
142	97 (R)	96 (R)	94 (R)	90 (R)	[51]
143	92 (R)	89 (R)	90 (R)	90 (R)	[50]

142 (ligand structure: 1,2-bis[(2-methoxyphenyl)(phenyl)phosphino]ethane)

143 (ligand structure: 1,2-bis[(2-methylphenyl)(phenyl)phosphino]ethane)

(Continued)

Table 1.4 (Continued) Results of asymmetric hydrogenations (ee [configuration of the product]) catalyzed by the Rh complexes of diphosphines **142–146**

Ligand	Ph—CH=C(NHCOMe)CO₂Me	Ph—CH=C(NHCOMe)CO₂H	AcO/MeO-aryl substrate	CH₂=C(NHCOMe)CO₂H	Refs.
144	97 (R)	90 (R)	91 (R)	93 (R)	[50]
145	>99 (R)	92 (R)	91 (R)	96 (R)	[50]

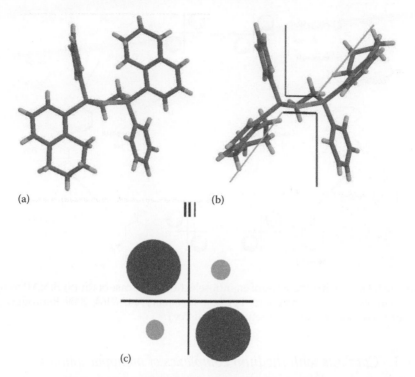

Figure 1.20 Models built with using the X-ray structure of the cyclooctadienyl-rhodium complex of the diphosphine **17**: (a) X-ray structure without the cod ligand; (b) structure obtained by rotation around P–C bonds of the tetrahydronaphtalenyl substituent; and (c) quadrant diagram corresponding to the upper structure—hindered quadrants are made by axial phenyls. (Reprinted with permission from Gridnev, I. D. and Imamoto, T., *Acc. Chem. Res.*, 37, 633–644. Copyright 2004 American Chemical Society.)

The sense of enantioselection in other reactions can be also analyzed using the conclusions made above. Thus, the structurally rigid Rh complex of (*R*, *R*)-QuinoxP* ligand[52] (**147**) always has the bulky *tert*-butyl substituent in the upper left quadrant, and the sense of enantioselection of asymmetric hydrogenation (stereoselection in octahedral Rh(III) complexes) is consistent with that of asymmetric addition of arylboronic acids to enones (stereoselection in square planar Rh(I) complexes) (Scheme 1.35).[52]

In contrast, as shown above, the asymmetric environment around the Rh atom changes when the geometry of the Rh–(*S*)-BINAP complex (**148**) transforms from octahedral to square planar. Accordingly, whereas the sense of enantioselection of asymmetric hydrogenation catalyzed by **147** is the same as in the case of **148**,[53] an opposite sense of enantioselection is observed in the asymmetric addition of phenylboronic acid to cyclic enones catalyzed by **147** and **148**.[54]

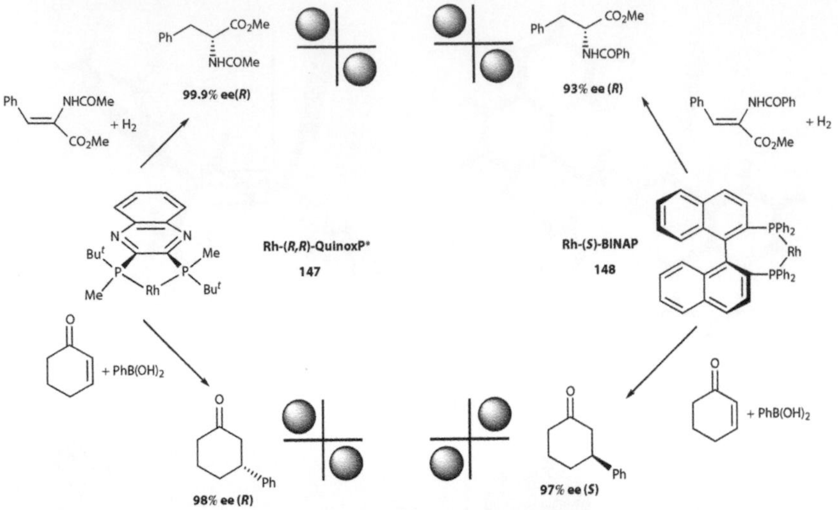

Enantioselection in asymmetric catalysis

Scheme 1.35 Switch of the sense of enantioselection in the case of Rh-(S)-BINAP complex. (Gridnev, I. D. and Imamoto, T., *Chem. Commun.*, 7447–7464, 2009. Reproduced by permission of The Royal Society of Chemistry.)

1.1.3 Catalysis with rhodium complexes of monophosphines

Catalytic systems employing Rh complexes of chiral monophosphines are quite popular recently due to the higher accessibility of the monophosphorus ligands, which makes them amenable to high throughput screening techniques.[55–57] There are two main problem of the mechanistic studies in this field. The first is the ambiguity in the actual composition of the active catalyst, since any of the species RhL, RhL$_2$, RhL$_3$, and RhL$_4$ are available in the reaction mixture.[55] Initially applied nondirect approaches to this problem provided contradictory information. Based on the observation of nonlinear effects[55,58] it was suggested that RhL$_2$ species must be responsible for the enantioselective hydrogenation rather than RhL species, as has been suggested in another work.[59] Another problem is the conformational flexibility of the complexes with monodentate ligands that makes ambiguous any straightforward stereochemical correlations.

Further research showed convincingly that the RhL$_2$ species are likely to be responsible for the enantioselective catalysis that makes the mechanisms of the asymmetric hydrogenation catalyzed by Rh complexes with mono- or diphosphine ligands similar.

Thus, clean formation of the solvate complex [Rh(L)$_2$(S)$_2$]$^+$ (**150**) where L = *t*-BuP((S)-binaphthoxo) existing in equilibrium with its dimer **150a** (Scheme 1.36) was reported.[60] Only very weak binding with MAC (**4**) or benzoyl phosphonate (**73**) was detected at −100°C. Nevertheless, using ^{13}C

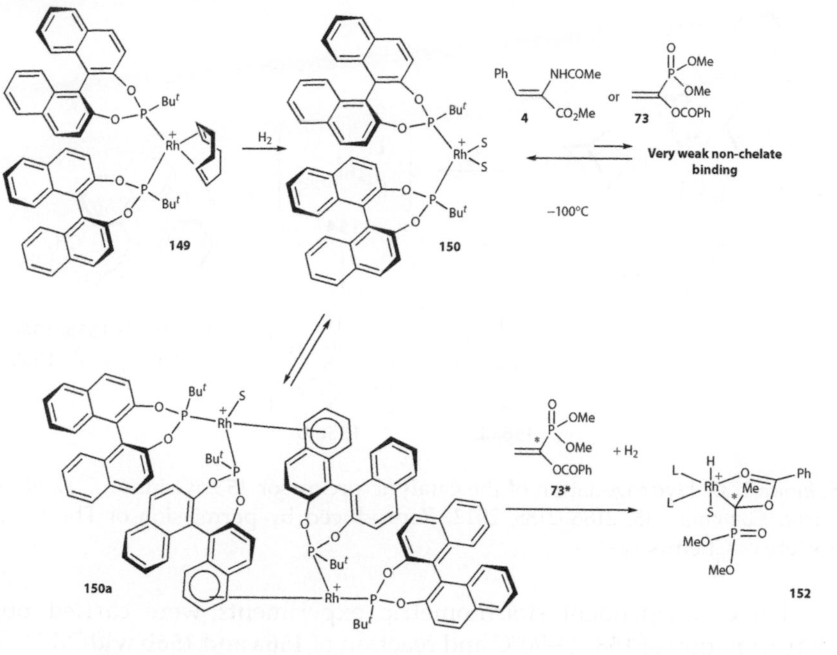

Scheme 1.36 Intermediates detected with the bis-monophosphorus–Rh complex 71 as a catalyst. (Gridnev, I. D. and Imamoto, T., *Chem. Commun.*, 7447–7464, 2009. Reproduced by permission of The Royal Society of Chemistry.)

labeled phosphonate **73*** and hydrogenating it in the presence of catalyst it was possible to characterize the monohydride intermediate **152**. Two mono-phosphorus ligands are present in the structure of **152**, hence the other intermediates in the catalytic cycle are likely to be RhL_2 species either.[60]

Hydrogenation of the catalytic precursor [Rh(NBD) ((*R*)-PhenylBinepine)$_2$]SbF$_6$ (**153**) in donating solvents like CD$_3$OD or THF-d$_8$ at ambient temperature resulted in quantitative formation of the corresponding solvate dihydrides **156a** or **156b**, respectively with the *cis*(H)-*trans*(P) structure.[61] Apparently, *cis*(H)-*cis*(P) dihydrides **155a, b** initially formed *via* oxidative addition, further rearranged into the more stable compounds **156a, b** (Scheme 1.37).

Computations confirmed that **155a** is 6–8 kcal/mol less stable than **156a** (depending on the conformation), hence the former cannot be detected in the NMR spectra, but is nevertheless kinetically accessible.

In less polar CD$_2$Cl$_2$ hydrogenation stopped at the stage of the solvate complex **154** which reversibly formed dimer **157** and gave diastereomeric chelate catalyst–substrate complexes *re*-**158** and *si*-**158** in a 4:1 ratio upon addition of MAC.[61]

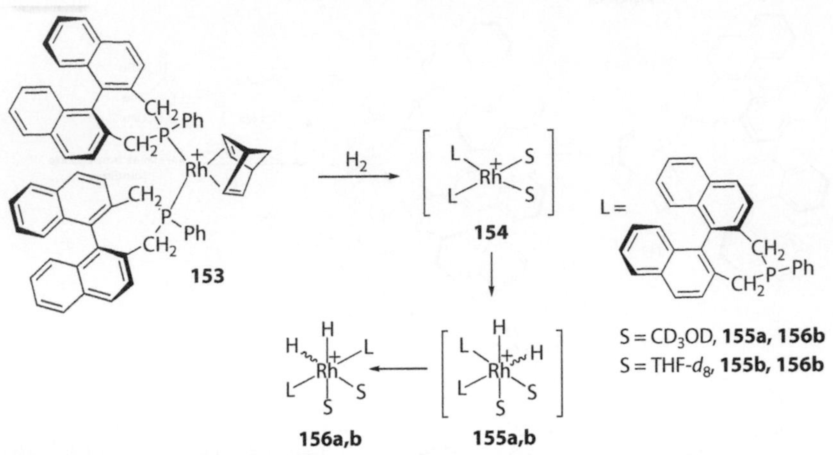

Scheme 1.37 Hydrogenation of the catalytic precursor **153**. (Gridnev, I. D. et al., *Chem. Commun.*, 48, 2186–2188, 2012. Reproduced by permission of The Royal Society of Chemistry.)

Three independent stoichiometric experiments were carried out: hydrogenation of **158** at −90°C and reaction of **156a** and **156b** with MAC at low temperatures. Despite the difference in the experimental set-ups and in the temperature regimes of hydrogenation, the product **72** obtained in these experiments was always of >99% ee (*S*) (Scheme 1.38).

If the reaction product could be obtained directly via oxidative addition of H_2 to *si*-**158** and *re*-**158**, then their hydrogenation at decreased

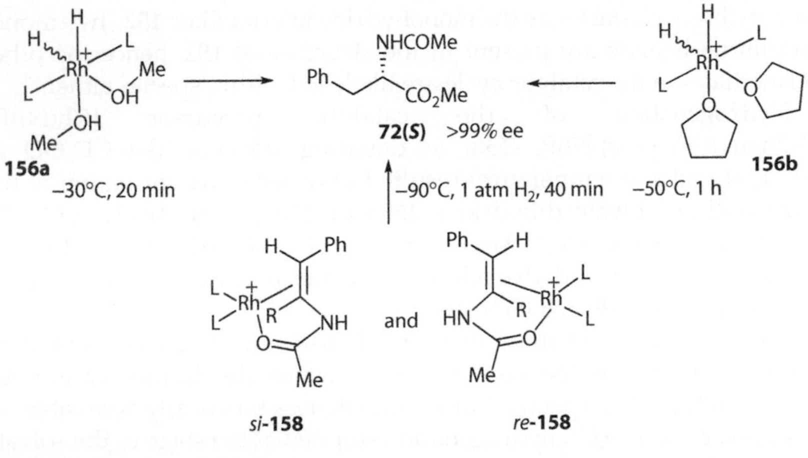

Scheme 1.38 Three different experiments carried out at decreased temperatures. (Gridnev, I. D. et al., *Chem. Commun.*, 48, 2186–2188, 2012. Reproduced by permission of The Royal Society of Chemistry.)

temperature, under the conditions of the slow exchange between *si*-**158** and *re*-**158**, would inevitably result in the decrease of the optical yield of **72**. Hence, either the double bond is dissociated during or after the oxidative addition of H_2 to *si*-**158** and *re*-**158**, or the oxidative addition itself occurs indirectly via partial or complete dissociation of the substrate. In other words, immediately after the hydrogen activation the double bond of the substrate is not coordinated, and the reaction pool contains various nonchelating intermediates rapidly interconvert via intra- and intermolecular coordination-decoordination processes, oxidative additions and reductive eliminations of H_2 (Scheme 1.39).[61]

Therefore, the stereoselection occurs on a later stage of the catalytic cycle, after selective coordination of the double bond with one of its prochiral planes to the octahedral Rh(III) complex that already contains activated hydrogen. The migratory insertion step is characterized by very low activation barriers, hence it takes place immediately after the double bond acquires the appropriate conformation coplanar with the P^trans–Rh–H bond.

Remarkably, the reversibility of the reaction is conserved until the stage of the monohydride intermediate **159** (Scheme 1.39).[61] Reversibility of all stages (and even of the stage in which the optical center is being created) of the catalytic cycle preceding the irreversible release of the product and regeneration of the catalyst effectively levels the effect of the multiple reaction pathways and excludes the possibility of a racemization of the already created asymmetric center.

The outlined mechanism is in a complete accord with the stereoselection mechanism suggested previously for the Rh-diphospine catalyzed hydrogenations (*vide supra*).

Due to the conformational flexibility of the Rh complexes with monodentate ligands, computational studies of the enantioselective step are complicated, and such studies were not reported so far. Nevertheless it was shown experimentally and computationally that distinct conformational minima can be observed for RhL_2^+ species that opens further possibilities for the computational analysis of these systems.[62]

1.1.4   Conclusions

A relatively large amount of experimental and computational data acquired so far for the Rh-catalyzed asymmetric hydrogenation makes possible some definite conclusions on the nature of the catalytic cycle and the mechanism of enantioselection that often result in the ultimate catalytic performance.

The first important feature of the catalytic cycle is its reversibility expanded until the very late steps. Practically each step in a chiral catalytic cycle can demonstrate certain selectivity, and it is difficult to imagine a situation when each stereoselective step would contribute to the accumulation of the same handedness of the product. For example, it is clear from

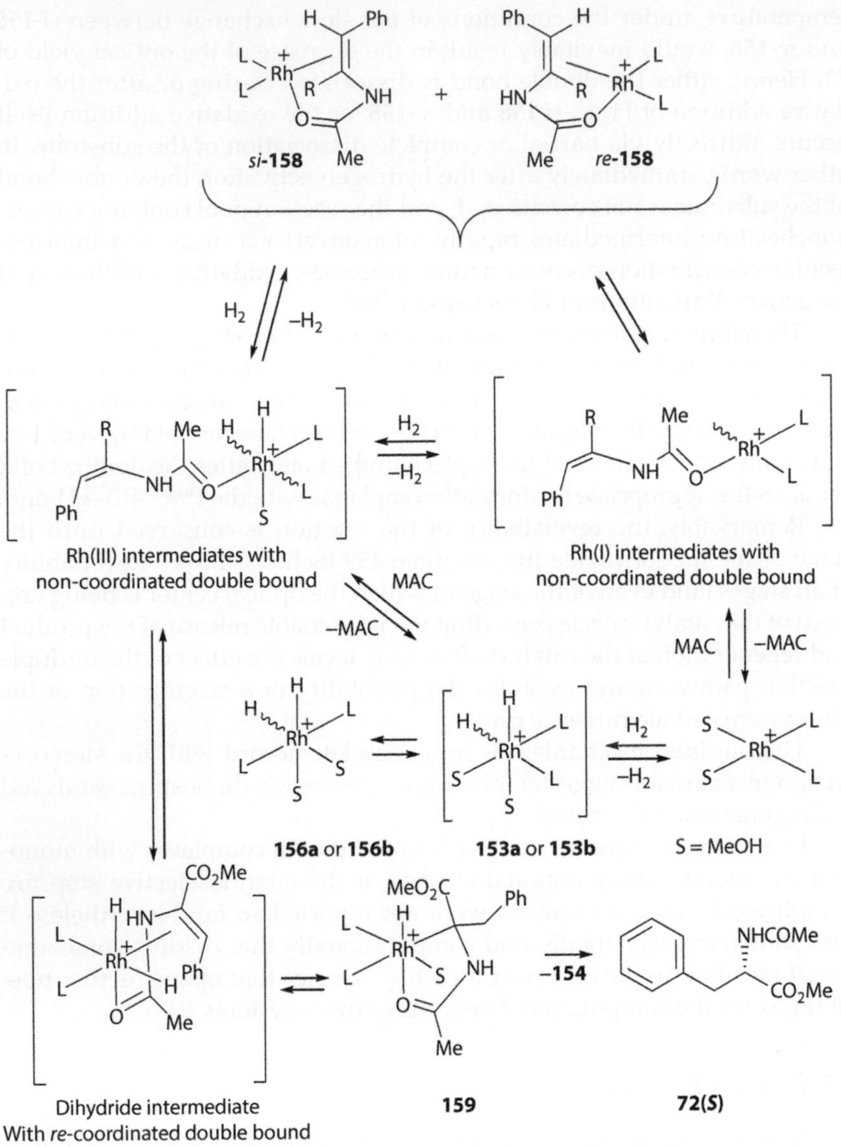

Scheme 1.39 Mechanism for the formation of the hydrogenation product **72(S)** in asymmetric hydrogenation of MAC catalyzed by the Rh complex of Binepin. (Gridnev, I. D. et al., *Chem. Commun.*, 48, 2186–2188, 2012. Reproduced by permission of The Royal Society of Chemistry.)

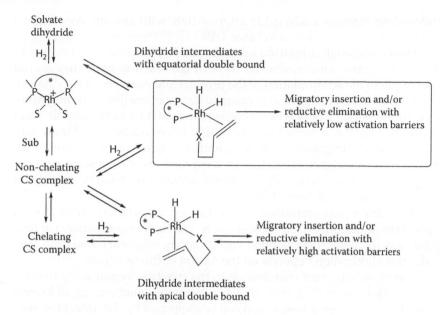

Scheme 1.40 The flux of catalysis can proceed only via the dihydride intermediates with equatorially coordinated double bond.

the data summarized in this chapter that the stereochemical preferences of the square planar Rh(I) complexes and octahedral Rh(III) complexes are opposite.

Due to the reversibility of all steps, even if initially a "wrong" pathway is taken, after encountering with a high activation barrier, reverse reaction can take place opening the door to a more facile reaction trail. This feature was called "trial-and-error" mechanism of enantioselection.[43]

This idea is illustrated in Scheme 1.40. Equatorial coordination of the double bond in the octahedral Rh(III) dihydride intermediate uniformly leads to the extremely facile migratory insertion and reductive elimination steps. The same reactions for the intermediates with apically coordinated double bonds are significantly less facile. Thus, the difference in free energies for the migratory insertion for the dihydride intermediates with apical and equatorial coordination of the double bond was computed to be 10–15 kcal/mol in the case of the asymmetric hydrogenation of various enamides[21] or α-dehydroamino acids.[16,22] In the case of the esters of β-dehydroamino acids this difference is smaller (3–7 kcal/mol, *vide supra* the Table 1.2), nevertheless in neither of computed catalytic cycles the pathways proceeding through apical double bond coordination actually competed with those proceeding through equatorial double bond coordination. Example of a catalytic cycle where the high barrier of reductive elimination makes impossible the realization of a pathway proceeding through a dihydride intermediate with apically coordinated double bond is described in Scheme 1.19.

Hence, although dihydride intermediates with apical coordination of the double bond can occasionally form in the reaction pool, they do not contribute to the accumulation of the product and recovery of the catalyst.

Furthermore, numerous experimental data confirm that facile migratory insertion is possible only when a coplanar orientation of the double bond and P–Rh–H^{trans} is achieved in a dihydride intermediate (e.g., Figure 1.4). Apparently, the migratory insertions with alternative coordination of the double bond are higher in energy, but computations of such steps were not tried so far. Nevertheless, only four possible pathways must be considered as competitive (e.g., Scheme 1.34).

The above considerations are valid for all known substrates of Rh-catalyzed asymmetric hydrogenation. The further competition of four pathways involving double bond coordination, migratory insertion, and reductive elimination depends on the nature of the substrate.

The most important contribution to the enantioselection is the requirement to form the chelate cycle in a less hindered quadrant. In all known cases, the sense of enantioselection can be explained by this rule. However, there are two ways to satisfy this requirement, and which one would be actually chosen depends on the properties of the substrate (e.g., Figure 1.15).

The same properties determine how many and which pathways would compete with the most facile one. There are cases when the most facile pathway meets almost no competition. Apart from the perfect ee's they can be recognized by the absence of the temperature effect on the optical yield, several such cases were described above. From the kinetic point of view they are characterized by undisturbed formation of the chelate cycle in the less hindered quadrant followed by barrierless and practically irreversible migratory insertion (e.g., red line in Figure 1.15a).

On the other hand, the competition of three pathways is clear from Figure 1.6 (Scheme 1.27). In general case, the accurate analysis of the experimental ee values must involve kinetic simulations of three steps: double bond coordination, migratory insertion and reductive elimination of two α-dihydride and two β-dihydride pathways.

1.2 Ru-catalyzed asymmetric hydrogenation and transfer hydrogenation of ketones

1.2.1 Overview

Ketones are among the most common unsaturated substrates containing a C=O group.[63] Homogeneous transition metal-catalyzed asymmetric hydrogenation (AH) and transfer hydrogenation (ATH) of prochiral ketones is one of the most powerful and efficient methods for the preparation of optically active alcohols.[63–86] The hydrogenation process uses cheap molecular hydrogen (H_2), the most abundant molecule in the universe and the most atom-efficient reducing chemical agent, and produces enantioenriched alcohols without forming any waste and with minimal workup involved.[87] If a hydrogen donor is different from H_2 (e.g., propan-2-ol, a HCO_2H/NEt_3 mixture, HCO_2Na/water etc.), the process is known as transfer hydrogenation.[64,82,88–90] The selectivity in terms of stereo-, chemo-, and regioselectivity could be different from AH systems; therefore, ATH may complement the latter. When hydrogenation is carried out in alcohols, these two processes may be concurrent, although hydrogenation is usually the dominant process. Beginning with a moderate enantiomeric excess (ee) of 82% reported by Markó et al. in 1985 (BDPP–Rh[I] complex),[91] the optical yields in the asymmetric hydrogenation of aromatic ketones in particular, have rapidly grown and achieved >99% ee. The most significant progress was made in the mid-1990s, when Noyori, Ikariya, and coworkers discovered the novel and very practical (pre)catalysts, *trans*-[$RuCl_2${(S)-binap}{(S, S)-dpen}][86,92–99] **1** and (S)-RuCl[(R, R)-XCH(Ph)CH(Ph)NH_2](η^6-arene) [X = NTs,[100–102] O[103]] **2** as shown in Chart 1.1, for the enantioselective hydrogenation and transfer hydrogenation of aromatic ketones, respectively.

The discovery of catalysts **1**[104–109] and **2**[110–130] contributed to the development of the asymmetric hydrogenation and transfer hydrogenation of ketones (and imines) into a key technology for small- to industrial-scale production of optically active compounds, including medicines, agrochemicals, and perfumes.[70,71,77,81,105,106,131–141] Although other practical catalysts were developed for ketone hydrogenations,[142–146] it is with catalysts bearing at least one N–H functionality,[147] that is, those for which **1** and **2** are prototypes, that enantioselective ketone hydrogenations have achieved the highest performance, closely approaching that of natural enzymatic systems and delivering, not infrequently, chiral alcohols with up to 99.9% ee with extremely small catalyst loadings (~10^{-5} mol %) at ambient temperatures making this reaction one of the most efficient

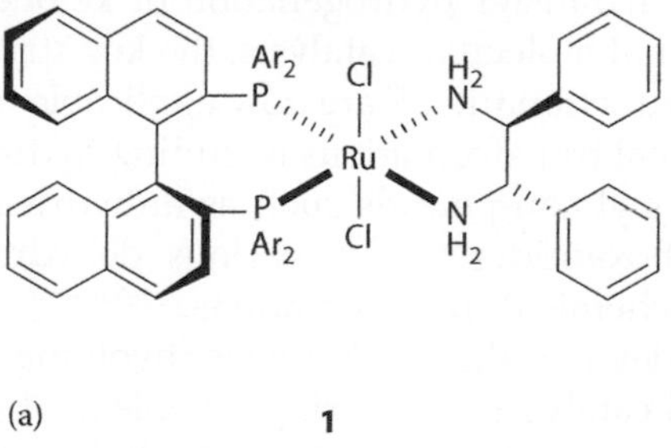

(a)   **1** 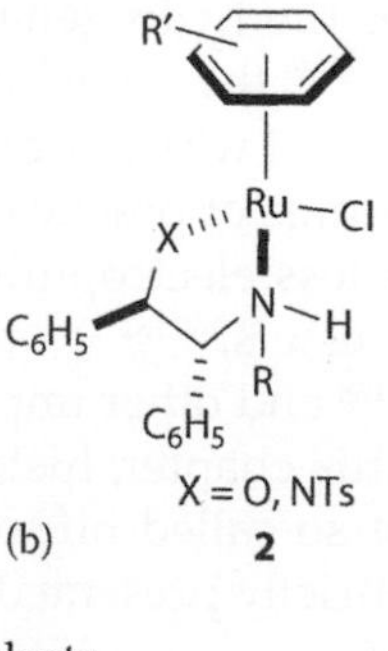   (b)   **2**

Chart 1.1 Noyori (1995) and Noyori–Ikariya (1995) catalysts.

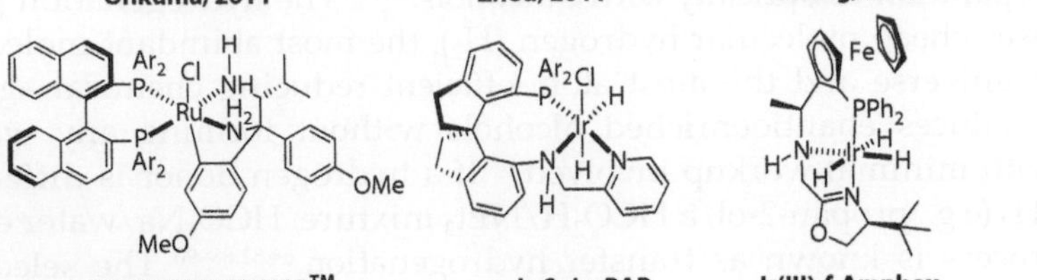

Ohkuma, 2011   Zhou, 2011   Zhang, 2015

(R)-RUCY-XylBINAP™   **Ir-SpiroPAP**   **Ir(III)-f-Amphox**

For acetophenone:	For acetophenone:	For acetophenone:
S/C = 100 000	S/C = 5 000 000	S/C = 1 000 000
solvent = EtOH/^{i}PrOH (1:1); 50 atm H$_2$	solvent = EtOH; 100–60 atm H$_2$	solvent = ^{i}PrOH; 80 atm H$_2$
11°C–35°C, 6 min	25°C–30°C, 15 days	48 hour
> 99% yield	91% yield	> 99% conversion
> 99% ee (S)	98% ee (S)	**up to 99.9% ee**

Chart 1.2 Practical hydrogenation catalysts for enantioselective ketones reduction.

artificial catalytic reactions developed to date. The most efficient and stable catalysts are typically based on Ru and Ir. An example of such a system is the latest modification of **1**, the chiral ruthenabicyclic complex (R)-RUCY-XylBINAP™ developed by Takasago Int. Corp. that quantitatively hydrogenates acetophenone into (S)-1-phenylethanol with >99% ee under 50 atm H$_2$ pressure within 6 min (11°C–35°C) with S/C = 10^5, Chart 1.2. The turnover frequency (TOF) reaches about 35,000 min^{-1} in the best case.[106,148] Another example of such a system is Zhou's chiral iridium catalyst Ir-SpiroPAP bearing a tridentate ligand with an N–H functionality that hydrogenates acetophenone with an S/C ratio up to 5 × 10^6 at 25°C–30°C producing the product in 91% yield and 98% ee, Chart 1.2.[149] Very recently, Zhang reported the novel air-stable chiral Ir catalyst Ir(III)-f-Amphox which hydrogenates aromatic ketones with up to 99.9% ee's, Chart 1.2.[150]

A deep understanding of the architecture of molecular catalysts **1** and **2** stimulated the development of the field known as "metal–ligand bifunctional catalysis" or simply "bifunctional catalysis"[74,76,83,151] that has been rapidly expanded upon in both academia and industry. Originally developed for the hydrogenation and transfer hydrogenation of ketones and imines,[152-159] the same bifunctional molecular catalysts, the key structural parameter of which is an N–H functionality,[147] are now applicable toward a wide range of practical chemical transformations including hydrogenations of less electrophilic carbonyl compounds such as anhydrides,[160,161] imides, esters,[160,162-164] and carboxamides,[160,164-166] various dehydrogenations,[167,168] and other important chemical transformations.[74,167,169,170]

In this chapter, historical views of the mechanisms involving prototypes of so-called bifunctional catalysts, pioneering complexes **1** and **2**, will be briefly presented and the current understanding will be critically assessed.

1.2.2   A brief critical overview of experimental and computational techniques used in the early mechanistic studies of 1 and 2

Mechanistic studies of Ru-catalyzed asymmetric hydrogenations and transfer hydrogenations began shortly after the discovery of catalysts **1** and **2** and were based on three main approaches: stoichiometric NMR studies, kinetic analyses including measurements of kinetic isotope effects (KIE's), and gas-phase computations. All of these methods are indeed ubiquitously used by chemists to study reaction mechanisms and analyze catalytic cycles.

1.2.2.1   The characteristic time and sensitivity of NMR spectroscopy

The characteristic time of a physical method derives from the uncertainty relation between energy and time.[171] Each act of interaction of an electromagnetic wave or particle flux with a substance occurs within a certain amount of time. If during this time the system rearranges, then the parameters of the system are averaging in time. In other words, if the lifetime of an intermediate is short compared to the characteristic time of the applied detection method used, it will not be observed by this method. The lack of detection of an intermediate by a specifically used detection method, however, often serves as "evidence" about the concertedness of a reaction involving the cleavage/formation of multiple bonds. Indeed, chemists usually intuitively define a *concerted multibond reaction*[172] (synchronous or asynchronous)[173,174] as one that takes place in a single kinetic step.[172,175–178] The widely used technique of NMR spectroscopy represents one of the slowest physical methods (time-scale of 10^{-7}–10^{-8} s) and therefore *cannot* be universally used to study *concerted multibond reactions*.[172] Possibly the best method to study these reactions, although less available than NMR spectroscopy, is femtosecond transition state spectroscopy, introduced in the 1990s by Zewail and coworkers, which has a time-scale of 10^{-15} s.[179] Time-resolved experiments have indeed unraveled the issue of concertedness for a variety of organic chemical reactions that were thought to be *concerted multibond*: α-cleavage of acetone, decarbonylation of cyclopentanone and the cracking of cyclobutane. These reactions were shown to be stepwise, proceeding through detectable high-energy radical intermediates.[180,181]

Finally, one should not forget about the sensitivity of the NMR method: if one observes, for example, two species by NMR, there are literally fifty two that he or she does observe by NMR.

1.2.2.2   Kinetics

Catalysis is a kinetic phenomenon.[182] Traditional kinetic analyses rely on the time-dependence of concentrations.[183] However, even for relatively simple (noncatalytic) reactions, commonly employed models for reaction kinetics such as transition state theory[184,185] (TST) and Rice-Ramsperger-Kassel-Marcus[186–189] (RKKM) theory, already being very rough approximations, cannot in many cases adequately describe chemical reactions.[190–192] The same is true for kinetic isotope effect measurements, which in addition can be obscured by H/D scrambling,[193] quantum tunneling effects[194] or posttransition-state bifurcations[195] on the potential energy surface (PES).[195–199] For a catalytic reaction that is composed of numerous steps, the interpretation of these data seems to be much more complicated: even the simplest catalytic cycles invoke kinetics that are more complex than can be easily handled by the usual tools for the extraction of rates.

Catalytic kinetics in the twentieth century were dominated by rate equations.[200] Rate constants, were and are, extracted from rate equations obtained by fitting kinetic data,[200] usually obtained by adjusting the process parameters to enable linearity.[201] A catalytic cycle, however, is a nonlinear dynamic system. Even with a fixed set of parameters, turn-over limiting states may change with time and extent of turnover.[201] Thus, depending on the portion of the catalytic reaction under study, the rate law may be different. Therefore, can statements as to the kinetic order of the overall catalytic reaction with respect to either substrate(s) consumption or product production obtained by traditional concentration kinetics always be universally assumed to be correct? *Even if they are, different reaction mechanisms may predict the same overall reaction rate.*

Boudart suggested that from an experimental standpoint, a better analysis of catalytic reactions is an iterative microkinetic analysis originally introduced in heterogeneous systems[202] based on rate constants of the assumed elementary steps, that would be calculated by computational chemistry in the future.[200] Chen used differential kinetics (time-dependence of rates rather than concentrations) and numerical integration of the rate equations to determine the elementary rate constants for the analysis of the catalytic cycle for acetophenone hydrogenation with Noyori's catalyst in the presence of a base.[201] It was concluded that what was termed as the "turn-over-limiting step" (i.e., dihydrogen cleavage or hydride transfer) changes during the course of reaction and this changeover depends on the degree of turnover and other parameters. This further confirms that statements such as *the reaction is first-order with respect to a component A* or *the reaction is zero-order with respect to a component B*, and so on, cannot be always universally applied.

The key concept of a rate-determining step in catalysis further deserves a separate comment. Very recently, the deeply rooted paradigm of this concept was suggested to be flawed.[203–208] With respect to a catalytic reaction, even under the assumption that a TST is valid, neither one reaction step, commonly abbreviated as "the slowest step of the reaction" nor its associated transition state, determines the efficiency of a catalytic cycle, that is, its TOF.[209-212] In other words, it is incorrect to segregate the step with the highest barrier from a computed catalytic cycle and compare the corresponding barrier with the experimentally determined one of a catalytic reaction. For example, Morris and coworkers follow the traditional approach and compare the computed activation barrier of H–H bond cleavage via metal–ligand cooperation with the experimentally determined activation barrier from kinetics studies in which the kinetic law is zero with respect to ketone, but first order in catalyst and H_2, that is, rate = k[catalyst] $[H_2]$.[213,214] Within the current level of understanding, the experimentally determined activation barrier should instead be compared to a computed energetic span,[203-208] within which an experimentally determined rate law, that is, the reactant orders, can be also realized.[215,216] The relationship between experimentally determined rate constants (the k-representation) and a computational study of a catalytic cycle (the E-representation) is discussed in the reviews[203-208] of Kozuch and Shaik. Because of the limitations and other factors discussed above, it is evident that this model will not always succeed in describing experimental kinetics, but at least may approximate it in certain cases.[215-217]

1.2.2.3   Computations in the gas-phase

Theoretical modeling is typically performed by organometallic chemists using the framework of pervasive density functional theory (DFT).[218,219] The vast majority of these quantum chemical calculations are traditionally performed in the gas-phase (vacuum) and quite frequently by using simplified molecular models. The most typical motivation for this is "to reduce computational time."

The behavior of chemical reactions occurring in solution, however, is largely dictated by solvent effects.[220] Solute-solvent interactions have dramatic effects on molecular structures, energies and properties[221-223] as well as on the outcome of the reaction, including the sense of enantioselection.[224,225] Within the gas-phase, reactions without charge separation or charge distribution are common (e.g., radical and pericyclic reactions), but in solution, most reactions do involve charge separation and charge distribution.[220] Solvent effects can be grouped into two distinct components.[220,226] *Nonspecific* solvation or *macrosolvation* describes interactions between solute molecules and the solvent polarization electric field (reaction field) around the solute in solution. *Specific* solvation or *microsolvation* is defined as the formation of kinetically stable complexes between the solute and solvent, originating from hydrogen-bonding, charge transfer or electron-pair donor-acceptor complexes and other weak chemical interactions. A computational description of *nonspecific* solvation can be achieved by treating the solvent as a continuous medium characterized by its macroscopic dielectric constant.[227-233] This gives rise to the continuum or implicit solvation model.[222,234] When the solvent and solute interact only slightly, the continuum model is an efficient tool for a proper description of chemical systems, accounting for the effects of solvation on a molecular structure, its energetics, and dynamics.[228,230] A computational description of *specific* solvation can be achieved via explicit inclusion of one or more solvent molecules into gas-phase calculations. This gives rise to the discrete or explicit solvation model.[51,55] The best approach to take into account both *nonspecific* and *specific* interactions is to compute the solute properties by including a few explicit solvent molecules in the continuum solvent, thus giving rise to a continuum/discrete solvation model.[230,235–239] Usually the inclusion of only a small number of solvent molecules is sufficient to fill most of the discrepancy between energy diagrams calculated in the gas-phase and in solution.[240–242]

On the other hand, it is well-known experimentally that even slight simplifications of catalyst structure can result in significant losses in activity, enantioface selectivity and other parameters. Therefore the *use of gas-phase computations or simplified models in order to* reduce computational time *should be avoided by all means*, since very different mechanisms may be in operation in the presence and absence of solvent and for different catalysts.[243,244]

1.2.3   Progress of the reaction mechanism with 1 and 2

Both **1** and **2** were found to be remarkable and unique in several aspects: they provide quantitative chemical yields within minutes to hours, enantioselectivities of up to 99%, high C=O/C=C chemoselectivities, and extremely high turnover efficiencies, notably for the hydrogenation process with S/C ratios of up to few millions at room temperature.[63] The last two aspects were particularly unusual: opposite C=C/C=O chemoselectivities and much lower turnover efficiencies were typically observed for hydrogenation catalysts that were accepted to operate via a classical inner-sphere mechanism,[245–248] that is, when preliminary substrate coordination to the metal is involved. The startling effect of diamines as ancillary ligands for ketone hydrogenations and transfer hydrogenations using such Ru[II] systems was observed when primary or secondary amines were present within the ancillary ligand backbone. Complexes containing chelating amines lacking N–H groups, for example, N,N,N',N'-tetramethylethylenediamine (TMEDA) or (R, R)-N(Ts)CH(Ph)CH(Ph) NMe$_2$ were observed to be totally or *almost* inactive,[63,92,101] suggesting the importance of an N–H functionality for the efficacy and intimate nature of the catalytic reaction. All of these early observations promoted Noyori, Ikariya and coworkers to suggest a mechanistically novel, nonclassical "metal–ligand bifunctional mechanism" as the pathway responsible for the complete reduction of a carbonyl group.[74,76,82,83,86,93,95,96]

Nonclassical, because the reduction of a ketonic substrate takes place within the outer-sphere of the coordinatively saturated catalyst-complex, that is, without preliminary substrate coordination. This feature nicely

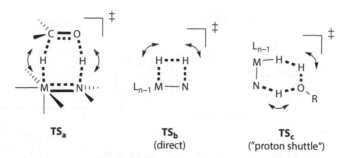

TS$_a$ **TS$_b$** **TS$_c$**
(direct) ("proton shuttle")

Chart 1.3 Transition states proposed by Noyori and co-workers.

rationalized the uncommon functional group selectivity in favor of the C=O moiety, as well as the unprecedented catalytic efficiencies. *Metal–ligand bifunctional*, since this outer-sphere reduction proceeds via a simple, aesthetically pleasing, *concerted* transfer of both the hydridic Ru–H from the metal and protic N–H from the ligand to a C=O linkage via a thermally allowed six-membered [σ2s + σ2s + π2s][249] pericyclic transition state[63,83,93–95,97] (TS) as shown in Chart 1.3, transition structure of type **TS$_a$**.

Within this framework, the nitrogen containing chelating ligand thus *directly participates* in the act of proton transfer (in concert with hydride transfer from the metal) via its N–H group, which is consistent with the observation that a chelating diamine with at least one N–H functionality is needed for activity.[250] The active form of the catalyst was proposed to be regenerated by reaction of the 16e⁻ Ru amido complex produced by this concerted transformation with molecular hydrogen or propan-2-ol, depending on whether the process is hydrogenation or transfer hydrogenation. For hydrogenation, this occurs either directly via *multibond concerted*[172] **TS$_b$**, or with the help of one protic molecule via a so-called "proton shuttle" type, *multibond concerted*[172] **TS$_c$**, respectively.[93] Here again, the H–H bond cleavage step proceeds via direct participation of the ligand in *multibond* breaking/making processes and within the currently accepted and popular point of view, is said to occur via "metal–ligand cooperation."[251,252] For transfer hydrogenation, the catalyst is regenerated via a transition structure of type **TS$_a$**. The corresponding catalytic cycles proposed from early studies for the active forms of catalysts **1** and **2** are shown in Schemes 1.41 and 1.42, respectively. They were discussed elsewhere.[63,83,93–95,97,253,254]

Here, are recalled several important features of these catalytic cycles gleaned from early studies. First, during the course of the reaction involving both catalysts, the oxidation state of the metal does not change, and the ligand undergoes a reversible chemical transformation that is N–H bond cleavage/formation. Second, all of the processes represented by transition states **TS$_1$**–**TS$_5$** are *multibond concerted*,[172] that is, both reduction of the substrate and catalyst regeneration via H–H bond cleavage involves

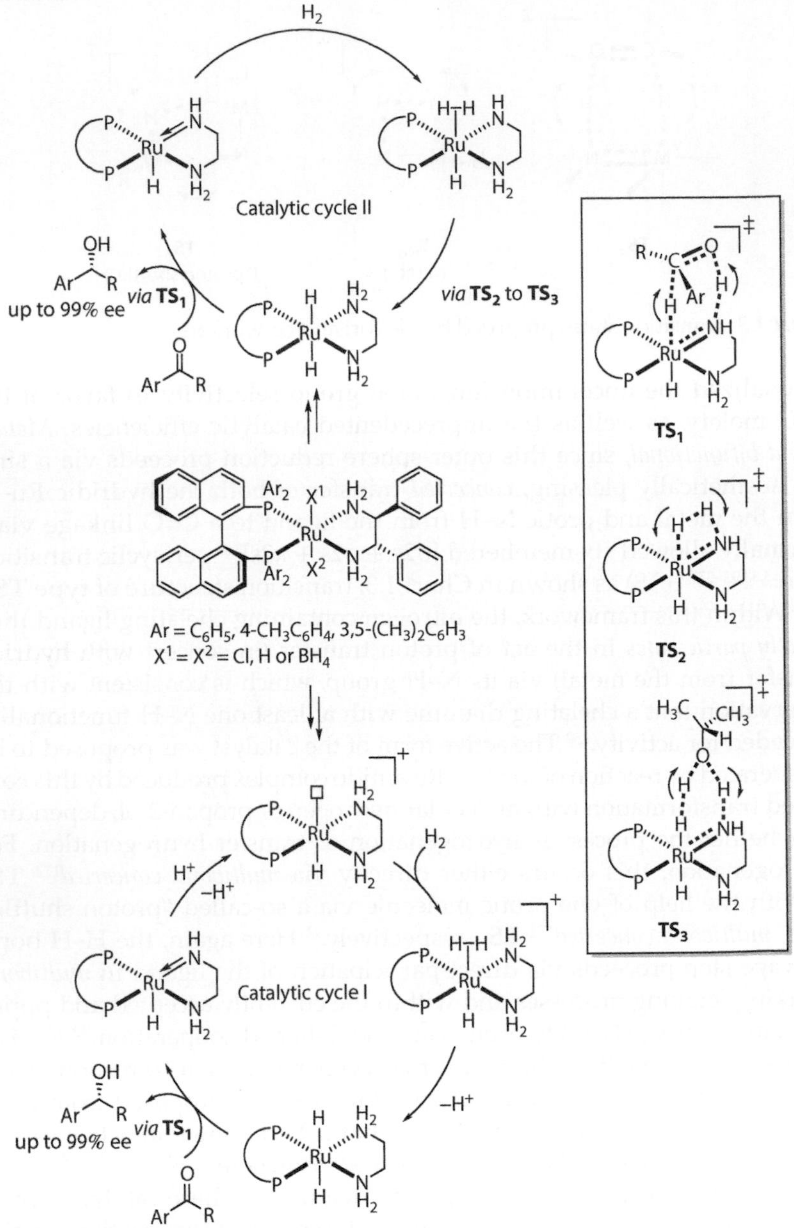

Scheme 1.41 Catalytic cycle for the hydrogenation of aromatic ketones by an active form of the Noyori (pre)catalyst **1** from early studies: catalytic cycle I (base-free conditions), catalytic cycle II (under high KO-*t*-C$_4$H$_9$ concentration). Formation of the major enantiomeric product is shown. (Reprinted with permission from Dub, P. A. et al., *Dalton Trans.*, 45, 6756–6781. Copyright 2016 Royal Society of Chemistry.)

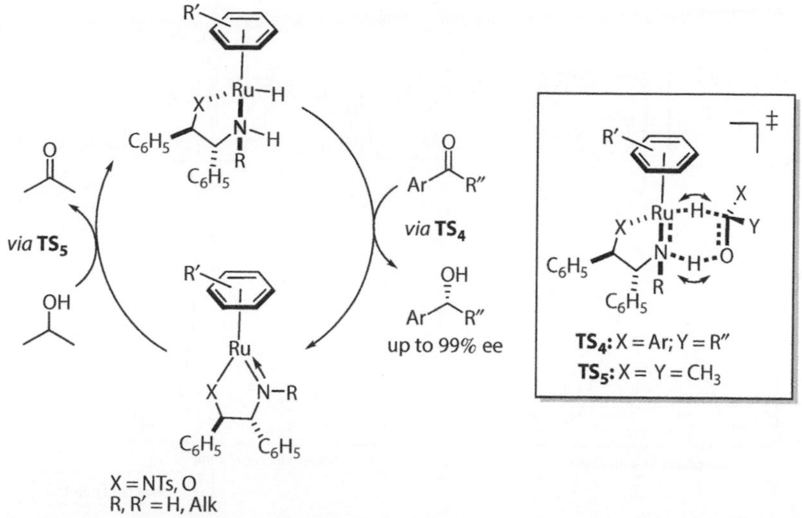

Scheme 1.42 Catalytic cycle for the transfer hydrogenation of aromatic ketones by an active form of the Noyori–Ikariya (pre)catalyst **2** from early studies. Formation of the major enantiomeric product is shown. (Reprinted with permission from Dub, P. A. et al., *Dalton Trans.*, 45, 6756–6781. Copyright 2016 Royal Society of Chemistry.)

the breaking and formation of multiple bonds in one single act. Third, the N–H proton in the 18e⁻ hydride complex plays a key role in hydrogen delivery to the ketone, while the amide nitrogen in the 16e⁻ amido complex cleaves H_2 or propan-2-ol. In other words, in both requisite parts of the catalytic cycle, that is, complete substrate reduction and catalyst regeneration, the ligand is "chemically noninnocent."[255–259] Catalytic cycles implied from early studies did not contradict available experimental data for **1**[260–262] and **2**[263–264] and were also supported by numerous gas-phase computations for **1**[214,249,265–279] and **2**,[280–286] respectively where saddle-points corresponding to **TS₁**–**TS₅** were indeed located. In this gas-phase profile, in comparing **TS₂** and **TS₃**, a "proton shuttle" type *multibond concerted* **TS₃** was found to be 12.7 kcal/mol more favorable.[279]

It has been recently found, however, that there are a lot of experimental and theoretical inconsistencies with some of these elementary steps, as well as for the whole catalytic cycle involving (pre)catalysts **1**[201,253,287–293] and **2**.[193,254,294–300] The most important realization came from computational chemistry incorporating solvent effects: on the contrary to the gas-phase, transition states **TSₐ** and **TS_c** corresponding to *multibond concerted* processes do not physically exist in solution. The curvature of the PES from gas to solution changes as follows: corresponding first order saddle-point

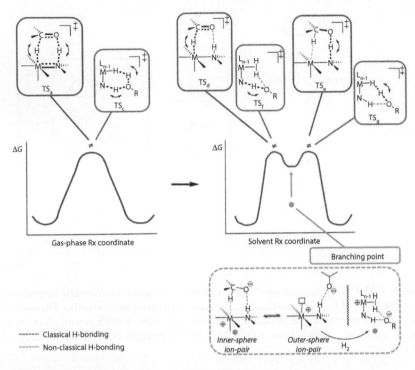

Figure 1.21 Schematic two-dimensional (2D) energy/reaction coordinate type diagram in gas phase and solution for the processes of complete C=O group hydrogenation/dehydrogenation and H–H bond cleavage/formation, respectively. The free energy (y-axis) is not scaled.

($\neq$) dissects into one stationary ($\bullet$) and two first order saddle-points ($\neq$) as shown in Figure 1.21.[253,254]

This means that in solution, the 1:1 reaction between 18e metal hydride complex and ketonic substrate takes place via a consecutive outer-sphere hydride transfer through transition state structure **TS$_d$** in the first step, and proton transfer from the ligand via transition state structure **TS$_e$** in the second step, respectively. The reaction proceeds through a well-defined high-energy intermediate likely of a short lifetime, which according to the classification of Macchioni,[301] is an inner-sphere ion pair. In addition to Coulomb interaction, this ion-pair is stabilized by two non-covalent-interactions, nonclassical C–H...M hydrogen bonding[302–304] and classical ionic N–H...O hydrogen bonding, respectively. For the active form of (pre)catalyst **1**, the presence of this high-energy inner-sphere ion-pair intermediate on the PES and its equilibrium with the outer-sphere ion-pair (Figure 1.21),[253] allows for the formation of either an alkoxo complex via O-coordination as was observed with ¹H NMR low-temperature

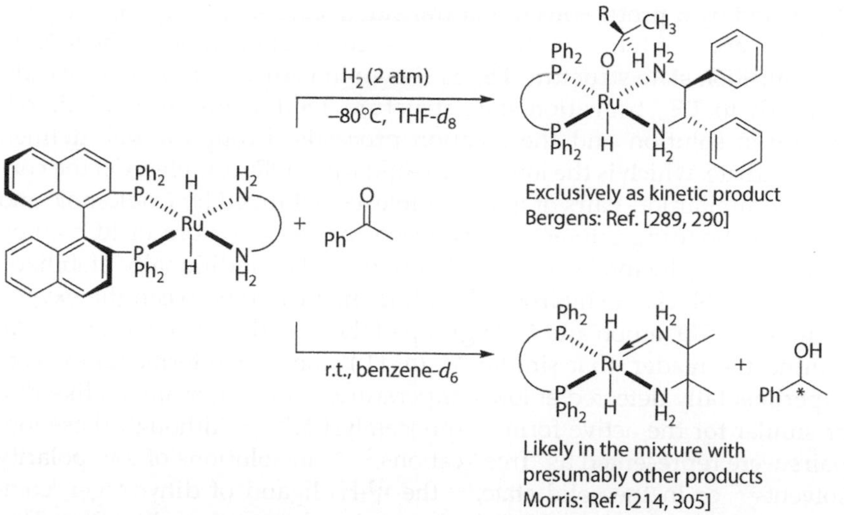

Scheme 1.43 Reported reaction between a trans-dihydride–Ru complex and acetophenone depending on the nature of the NN ligand and other parameters as determined by ^{1}H NMR experiment. The outcome of both reactions can be explained based on the existence of a high-energy and/or short lifetime intermediate, not detected (observed) by ^{1}H NMR spectroscopy. (Reprinted with permission from Dub, P. A. et al., *Dalton Trans.*, 45, 6756–6781. Copyright 2016 Royal Society of Chemistry.)

experiment by Bergen's group[289,290] or the formation of the 16e$^-$ Ru amido complex and 1-phenylethanol via N–H proton transfer from the ligand, likely in the mixture with other products as was observed with a ^{1}H NMR room temperature experiment by Morris' group with a complex structurally similar to Noyori's catalyst, *trans*-[RuH$_2$[(R)-binap](tmen)][214,305] as shown in Scheme 1.43, respectively. The identity of the products depends on the nature of the catalyst, temperature, solvent, recorded time after mixing the reagents, and so on.[253] Both these alkoxo and 16e$^-$ Ru amido complexes, however, were recognized as not being intermediates within the catalytic cycle, but rather as off-loop species.[253]

Similar experimental observations were made for the active form of (pre)catalyst **2**,[264,306] and they are all in agreement that the reaction between the 18e$^-$ Ru hydride complex and a ketone or the opposite reaction between the 16e$^-$ Ru amido complex and alcohol takes place via a high-energy and/or short-lived intermediate,[254] whose existence is in agreement with an ^{1}H NMR kinetics study.[264]

In a similar fashion, so-called "solvent assisted H–H bond cleavage via metal–ligand cooperation" (**TS$_c$** in Figure 1.21) that regenerates the catalyst represents a two-step process in solution: first N-protonation of the ligand by a protic solvent via transition state structure TS_f and then deprotonation of the η^2-H_2 ligand by the conjugated anion of the solvent via transition state structure TS_g as shown in Figure 1.21. In other words, similarly to TS_a, transition state structure TS_c has no mechanistic relevance in solution and the reaction proceeds through a well-defined intermediate, which is the ion-pair M–(η^2-H_2)$^+$ ... OR$^-$ complex. On the contrary to similar ion-pairs of η^2-H_2 complexes stabilized by Coulombic and hydrogen bonding interactions between the oxygen atom of RO$^-$ anion and the η^2-H_2 ligand,[303] such an ion-pair is also additionally stabilized by an ionic N–H ... O hydrogen bonding interaction between the oxygen atom of the RO$^-$ anion and N–H group of the ligand, respectively.[253,307] We remind the reader, that similar M–(η^2-H_2)$^+$ ion-pair intermediates were experimentally detected at low temperatures for counter anions like BF_4^- or similar for the active form of (pre)catalyst **1**.[291,292] Although these ion-pairs were represented as "free" cations,[291,292] in solutions of low-polarity solvents[303] or in the solid state,[308] the η^2-H_2 ligand of dihydrogen complexes was shown to establish hydrogen bonds to anions like BF_4^-. This likely explains the reported[291] unexpected kinetic acidity of NH protons versus the η^2-H_2 ligand in the cationic η^2-H_2 complex derivative of **1** with a BF_4^- counteranion: indeed the acidity of some cationic η^2-H_2 compounds is sometimes as strong as that of sulfuric or triflic acid.[309–311]

The existence of well-defined intermediates on a PES, which is actually multidimensional in reality, creates alternative catalytic pathways or results in their crossover. These alternative pathways proceed through steps in which the N–H ligand remains chemically intact, are not only energetically feasible, but could be dominant or almost exclusive pathways to reduce the substrate and/or regenerate the catalyst.[253,254] This leads to a very important conclusion: the N–H group of the bifunctional (pre)catalysts **1** and **2** *does not necessarily have to* participate in the bond cleavage/formation events through the course of the catalytic reaction.

For example, the microscopic reverse[312–314] for the deprotonation of an η^2-H_2 ligand by an RO$^-$ anion via TS_g represents protonation of neutral transition-metal hydride complexes by proton donors, a well-known synthetic procedure used to generate cationic (η^2–H_2)-complexes through dihydrogen-bonding (DHB) adducts.[303,309–311,315–320] For catalysts traditionally denoted as bifunctional such as **1**, such DHB adducts are additionally stabilized[253,307] by hydrogen-bonding with the N–H functionality as shown in Scheme 1.44a.

This additional interaction increases the strength of the primary DHB-bonding interaction between an alcoholic substrate and a catalyst nonadditively via the so-called cooperative effect in hydrogen bonding,[321] and could explain the well-known catalytic efficiency of bifunctional catalysts in the dehydrogenation of alcoholic substrates[167] as an alternative path to H–H bond formation via metal–ligand cooperation as shown

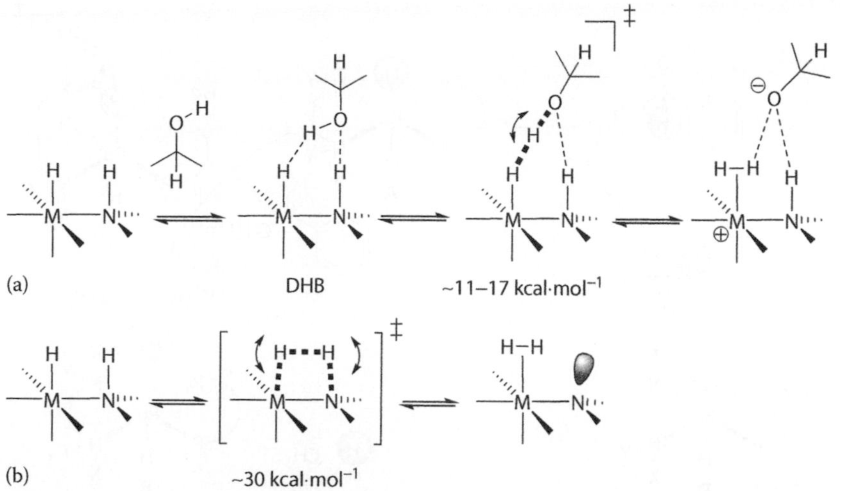

Scheme 1.44 Two ways of H-H bond formation. (Reprinted with permission from Dub, P. A. et al., *Dalton Trans.*, 45, 6756–6781. Copyright 2016 Royal Society of Chemistry.)

in Scheme 1.44b. Indeed, the computed activation barrier for the former is ~11–17 kcal/mol depending on acidity of the alcohol and the nature of the catalyst,[253,307] whereas for the latter case is ~30 kcal/mol![253] Therefore, the H–H bond is unlikely to be formed via metal–ligand cooperation.

The revised catalytic cycles for (pre)catalysts **1** and **2** identified based on a computed[253,254] minimum energy path (MEP)[195–199] are shown in Schemes 1.45 and 1.46, respectively. The details were discussed elsewhere.[253,254]

Similarly to previous views, the oxidation state of the metal does not change during the catalytic reaction and the ketonic substrate is indeed reduced within the outer-sphere, that is, without preliminary substrate coordination. However, this reduction proceeds via a *one-bond* type hydride transfer (H⁻) to afford the ion-pair intermediate in the first step (complexes **B** and **b** in Schemes 1.45 and 1.46, for **1** and **2**, respectively). The corresponding transition state of type **TS$_d$** was found to be enantio- and typically rate-determining.[253,254]

For the mechanistically more simple transfer hydrogenation process with **2** in particular, the charge-separated nature of this transition state persuasively explains the experimentally observed increase in the initial reaction rate and the average activity as a function of reaction medium polarity by Xiao and Liu[193] as well as Tanis.[299] Indeed such an observation is characteristic for so-called dipolar transition state reactions,[322,323] where activated complexes differ considerably in charge separation or charge distribution from the initial reactants, contrary to pericyclic reactions in

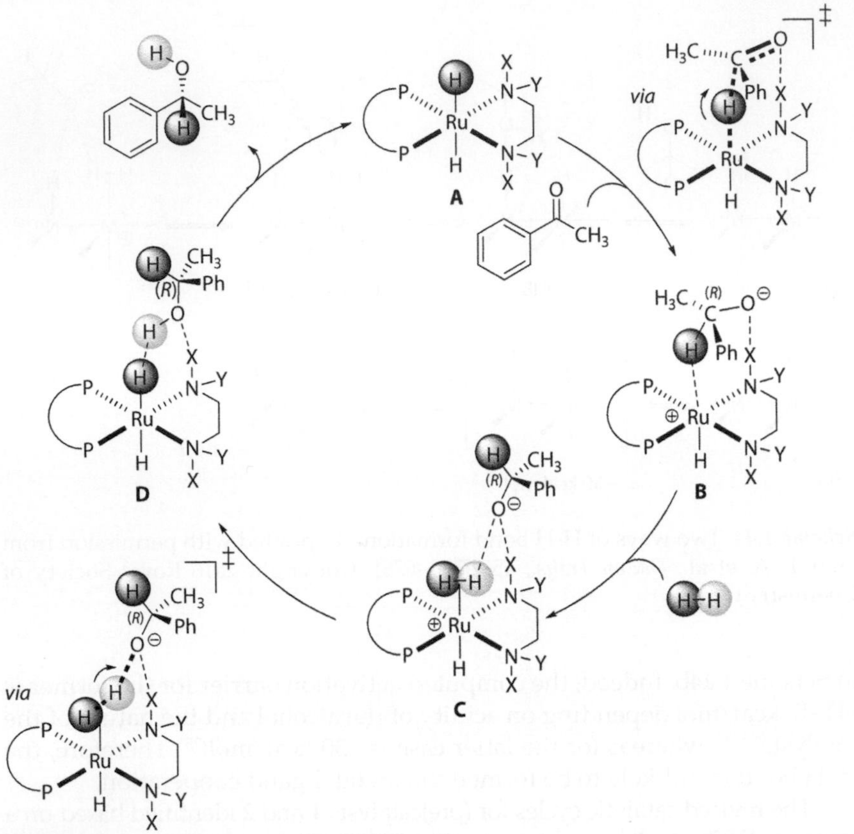

Scheme 1.45 A revised catalytic cycle for the asymmetric hydrogenation of aromatic ketones in propan-2-ol by Noyori's (pre)catalyst **1** based on a computed MEP[253] follows a H$^-$/H$^+$ outer sphere hydrogenation mechanism (see text). KO-*t*-C$_4$H$_9$-free conditions: X = Y = H. Under high KO-*t*-C$_4$H$_9$ concentration: X = Y = K and/or H. Formation of the major enantiomeric product is shown. (Adapted from Dub, P. A. et al., *J. Am. Chem. Soc.*, 136, 3505–3521. Copyright 2014 American Chemical Society.)

which the charge distribution within the activated complexes and the reactants is very similar (since no significant solvent polarity effects are expected). On the other hand, increasing the acidity of the N–H functionality will result in a stronger N–H...O hydrogen-bonding interaction, the key parameter that stabilizes the transition state of type **TS$_d$**, enabling faster catalysis, as indeed was observed with **2**.[294]

For the hydrogenation process with catalyst **1** in propan-2-ol, there are at least five possibilities for neutralization of the anion of the ion-pair product **B** obtained after this transformation and the regeneration of the

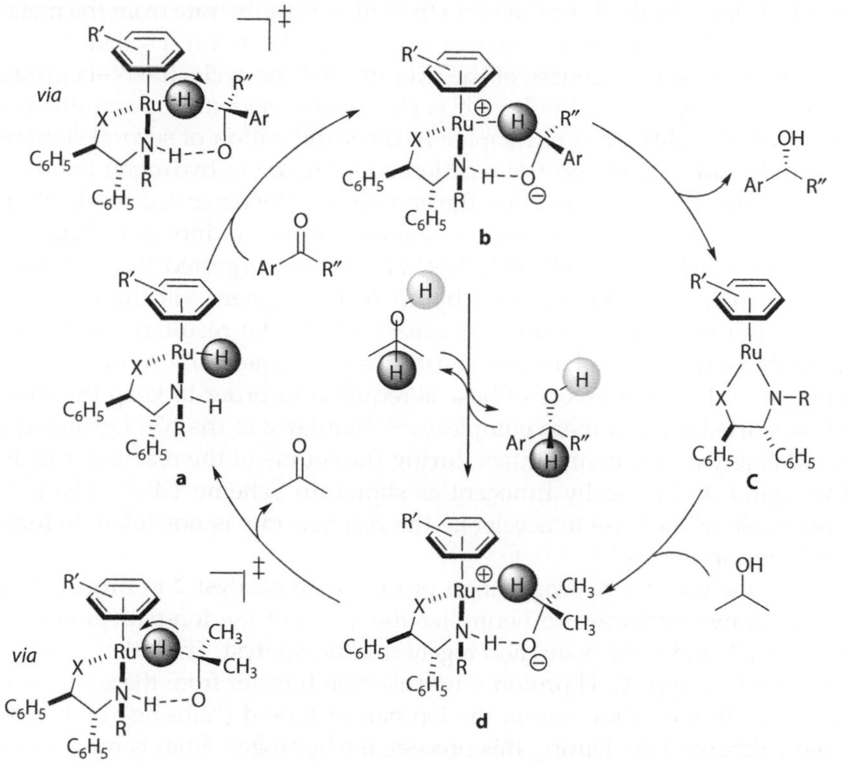

Scheme 1.46 A revised catalytic cycle for the asymmetric transfer hydrogenation of aromatic ketones in propan-2-ol by the Noyori–Ikariya (pre)catalyst **2** demonstrates crossover of the reaction pathways: the product is obtained via a H⁻/H⁺ outer-sphere hydrogenation mechanism and/or *step-wise* metal–ligand bifunctional mechanism (see text). Formation of the major enantiomeric product is shown. (Adapted from Dub, P. A. et al., *J. Am. Chem. Soc.*, 135, 2604–2619. Copyright 2013 American Chemical Society.)

catalyst via three associated TS's corresponding to H–H bond cleavage.[253] The MEP corresponds to catalyst regeneration via a *one-bond* type H–H cleavage by the same anion, that is, the conjugated anion of the not yet formed alcohol product. This is achieved via deprotonation of the η^2-H_2 ligand by the (*R*)-1-phenylethoxide anion in intermediate **C**, obtained from **B** after H_2 coordination as shown in Scheme 1.45. Therefore, the source of the proton for the reaction product is the η^2-H_2 ligand, and not the N–H functionality as was initially postulated within the framework of a metal–ligand bifunctional mechanism. We denote such a mechanism as an H⁻/H⁺ outer-sphere hydrogenation in order to state that the catalyst first transfers a hydride fragment to the unbound substrate from the metal center and then a proton is further transferred from a η^2-H_2 ligand.[324,325] Note that during the course of the overall catalytic cycle, the N–H group remains intact, that is, the ligand is *chemically innocent*. Importantly, the N–H functionality does participate in the stabilization of several stationary and saddle points on the reaction coordinate via hydrogen-bonding interactions. In order to explain the experimentally observed acceleration of the reaction rate in the presence of a large excess of inorganic base, for example, KOH, KO-i-C_3H_7 or KO-t-C_4H_9, it was suggested that the N–H functionality could be replaced by an N–K fragment via the chemical reaction of the catalyst with the base.[201,253,287,293] The resultant potassium amidato complexes reduce the substrates in a kinetically more efficient manner and a large excess of base is required in order to keep the equilibria shifted toward these complexes.[253] Similarly to the N–H group, the N–K moiety also remains intact during the course of the reaction, that is, the ligand is chemically innocent as shown in Scheme 1.45.[253] The proposed role of the base to accelerate the reaction rate is not solely to form amidato complexes.[253]

For the transfer hydrogenation process with catalyst **2** in propan-2-ol, there are two pathways to neutralize the anion of the ion-pair product **b** ("starting" branching point) and regenerate the catalyst. The latter proceeds via a *one-bond* type C–H proton + two electron transfer from the isopropoxide anion to the cation within the ion-pair of type **d** ("closing" branching point), Scheme 1.46. During this process, the hydrogen atom is repolarized as follows $H^+ \rightarrow H^-$. In one pathway, the source of the proton that neutralizes the anion in **b** is the ligand, thus the N–H ligand is *chemically noninnocent*. This mechanism for ketone reduction can be denoted as a *step-wise* metal–ligand bifunctional in order to state the difference with the conventional mechanism, which is concerted and realized only in the gas-phase. In the second path, the source of the proton used to neutralize the anion in **b** is propan-2-ol, therefore the N–H ligand *is chemically innocent* in this case. This is the same type of H^-/H^+ outer-sphere hydrogenation, except that the source of the proton used to neutralize the anion is now a protic solvent molecule. The crossover of these two reaction pathways is possible and likely takes place in all transfer hydrogenation processes catalyzed by bifunctional catalysts. Although, the relative contribution of each mechanism is computationally intractable,[254] the neutralization of the anion within ion-pair **b** by a protic solvent seems to be more probable, especially with increasing reaction medium polarity, for example, if the reaction is carried out in more polar formic acid or water.[254] In fact, Car–Parrinello molecular dynamics studies of Meijer[296] suggest that the catalytic reaction in water[80] proceeds exclusively or largely through, what we call here a H^-/H^+ outer-sphere hydrogenation mechanism.

1.2.4 The origin of the enantioselectivity

Stereoselection in the AH and ATH of aromatic ketones by the Noyori **1** and Noyori–Ikariya (pre)catalyst **2** takes place during the outer-sphere hydride transfer step, respectively.[253,254] The N–H functionality therefore not only stabilizes the corresponding transition state via ionic N–H...O hydrogen-bonding interaction, but also orients the outer-sphere position of the substrate along the Ru–H–C axis making stereodifferentiation possible.

For **1**, composed of two chiral ligands of the same absolute configuration (*S* and *S* or *R* and *R*), the extent of enantioselection typically increases in the row BINAP ≤ TolBINAP < XylBINAP,[326,327] that is, with increasing bulkiness of the PP ligand possessing axial chirality. Among NN ligands, DAIPEN is particularly effective.[327] For (pre)catalyst **2**, composed of one chiral ligand, better enantioselectivities are obtained for complexes based on *N*-sulfonylated 1,2-diamine, rather than amino alcohol ligands, leading to industrial use of the former.[82]

Typically ketonic substrates possessing electron-rich substituents (i.e., aromatic rings, alkynyl groups etc.), were identified to afford the reduced products with high enantiomeric excesses. Simple alkyl ketones, for example, are typically reduced with much lower ee's. Using absolute rate theory,[184,185] Figure 1.22 plots the energy difference ($\Delta G°_{298\,K}^{\neq}$) between averaged catalytic reaction activation barriers leading to major and minor enantiomeric products as a function of the experimentally observed enantioselectivity (ee, %) for the AH or ATH of acetophenone catalyzed by **1** and **2** or their derivatives, respectively.

For (pre)catalyst **1**, especially under high KO-*t*-C$_4$H$_9$ concentrations, the enantiodetermining step takes place most-likely on the (λ,λ)-conformer of the neutral Ru *trans*-dihydride complex and various isomers of mono- and di-amidato complexes through a set of isoenergetic transition states.[253] Although it was claimed that the enantioselectivity does not depend on the presence/absence of a large excess of inorganic base,[328] the base itself can catalyze the hydrogenation reaction to afford a racemic product.[329–331] All computational studies, no matter whether in continuum solvent[253] or in the gas-phase,[265,268,270,274,276] as well as experimental studies[260] are in agreement that the formation of the major product proceeds a through diastereomeric transition state that is stabilized either by N–H...π or N–K...π ligand–aromatic ring noncovalent attractive interactions in the catalyst–substrate complex and/or the absence of steric constraints as shown in Figure 1.23. In contrast, the transition state leading to the minor enantiomer is largely destabilized by the repulsive interaction between the aromatic ring of acetophenone and a portion of the binap ligand (phenyl substituent and one aromatic ring of naphthalene) as shown in Figure 1.23.

Computations suggest that the energy difference between the diastereomeric transition states increases proportionally to the number of

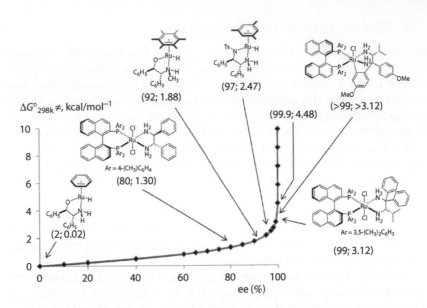

Figure 1.22 Energy differences between averaged catalytic reaction activation barriers leading to major and minor enantiomeric products as a function of the experimentally observed enantioselectivity for AH or ATH of acetophenone catalyzed by **1** and **2** or their derivatives, respectively. $RT = 0.59$ kcal/mol. (Reprinted with permission from Dub, P. A. et al., *Dalton Trans.*, 45, 6756–6781. Copyright 2016 Royal Society of Chemistry.)

K-atoms present in **1**. This was attributed to the more organized (rigid) structure of these complexes as a result of cation–π interactions.[253] Therefore, it is possible that an excess of inorganic base increases the enantioface-selectivity of the reaction via better stereo-differentiation within K-amidato complexes on one hand, while on the other hand, a background base-catalyzed reaction produces a racemic mixture. Due to competition of these two reactions, the observed enantioselectivity may appear not to depend on the presence/absence of a large excess of inorganic base.[328]

For the chiral ruthenabicyclic complex (S)-RUCY-XylBINAP™ developed by Takasago Int. Corp[106,148] the optimized geometries of the diastereomeric transition states leading to opposite enantiomers are shown in Figure 1.24. Their calculated energy difference is significantly higher than the one calculated[253] for complex *trans*-[RuH₂{(S)-BINAP}{(S, S)-DPEN}]. This is likely achieved due to much higher destabilization of the unfavorable transition state because of the much bulkier environment in this complex. We note to a reader that (S)-RUCY-XylBINAP™ affords products with much better enantioselectivities (>99% ee's).[106,148]

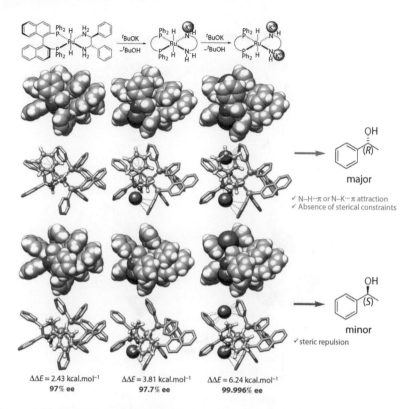

Figure 1.23 Transition states leading to *R*- and *S*-products: optimized[253] under DFT[218,219]/ωB97X-D[332]/SDD(Ru)/6-31G*(C, H,N, O,P, K)/SMD(propan-2-ol)[333] level of theory on the full model of *trans*-[RuH₂{(S)-BINAP}{(S, S)-DPEN}] and its K-amidato complexes with acetophenone. The enantioselectivity for each complex is calculated based on difference in electronic energy using absolute rate theory.[184,185] Some H-atoms are omitted for the ball and stick model. (Reprinted with permission from Dub, P. A. et al., *Dalton Trans.*, 45, 6756–6781. Copyright 2016 Royal Society of Chemistry.)

In the case of (pre)catalyst **2**, the enantioselectivity originates from the stabilization of the favorable diastereomeric transition state via multiple C–H...π attractive interactions[254,334] between the η⁶-ligand with the π-cloud of the approaching acetophenone, as well as from the destabilization of the unfavorable diastereomeric transition state via lone pair(s) – π repulsion[335] between oxygen (SO_2-) atoms within the Ts-ligand and the π-cloud of the approaching aromatic ketone as shown in Figure 1.25.[254]

Expanding on the results of Noyori's original computational work on the (pre)catalyst **2** based on an amino alcohol ligand,[282] in the case of the seminal catalysts based on *N*-sulfonylated 1,2-diamine ligands, a C–H...π interaction is still thought to be the main origin of enantioselectivity for the

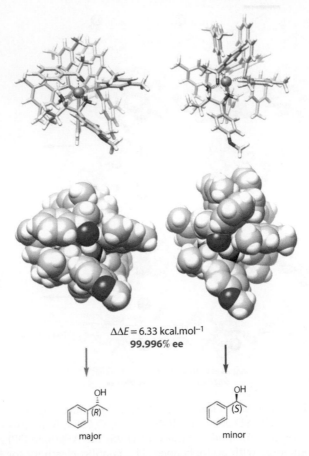

$$\Delta\Delta E = 6.33 \text{ kcal.mol}^{-1}$$
99.996% ee

major   minor

Figure 1.24 [RuH{(S)-xylbinap}{(S)-daipena}]–acetophenone: two diastereomeric transition states leading to (R)-1-phenylethanol and (S)-1-phenylethanol. Optimized under DFT[218,219]/ωB97X-D[332]/SDD(Ru)/6-31G*(C, H,N, O,P,)/SMD(propan-2-ol)[333] level of theory. The enantioselectivity is calculated based on difference in electronic energies by using absolute rate theory.[184,185]

latter.[252,336] However, it is essentially the presence of an SO_2 group and the corresponding destabilization of the minor diastereomeric transition state via lone pair(s) – π repulsion that makes catalysts based on N-sulfonylated 1,2-diamines highly enantioselective (90–99% ee) and significantly better than corresponding $Ru^{II}(\eta^6$-arene) complexes bearing chiral 2-amino alcohol auxiliaries (2–92% ee).[254] Therefore both *stabilization of the major* and *destabilization of the minor* diastereomeric TS's are important for achieving high ee. We note here that under certain conditions this lone pair(s) – π repulsion may become lone pair(s) – π attraction, Figure 1.25.[335] This could be the case for substrates such as 1,2,3,4,5-pentafluoroacetophenone and its derivatives that yield, as experimentally found, either racemates or even major

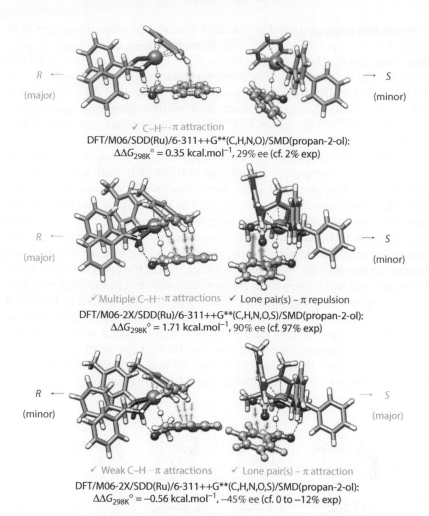

R ←
(major)

→ *S*
(minor)

✓ C–H···π attraction
DFT/M06/SDD(Ru)/6-311++G**(C,H,N,O)/SMD(propan-2-ol):
$\Delta\Delta G_{298K}° = 0.35$ kcal.mol^{-1}, 29% ee (cf. 2% exp)

R ←
(major)

→ *S*
(minor)

✓ Multiple C–H···π attractions   ✓ Lone pair(s) – π repulsion
DFT/M06-2X/SDD(Ru)/6-311++G**(C,H,N,O,S)/SMD(propan-2-ol):
$\Delta\Delta G_{298K}° = 1.71$ kcal.mol^{-1}, 90% ee (cf. 97% exp)

R ←
(minor)

→ *S*
(major)

✓ Weak C–H···π attractions   ✓ Lone pair(s) – π attraction
DFT/M06-2X/SDD(Ru)/6-311++G**(C,H,N,O,S)/SMD(propan-2-ol):
$\Delta\Delta G_{298K}° = -0.56$ kcal.mol^{-1}, –45% ee (cf. 0 to –12% exp)

Figure 1.25 (*S*)-RuH[(*R*, *R*)-N(Ts)CH(Ph)CH(Ph)NH$_2$](η^6-mes)–C$_6$H$_5$C(O)CH$_3$ or C$_6$F$_5$C(O)CH$_3$: optimized transition states leading to *R*- and *S*-products. The enantioselectivity is calculated[254] based on difference in free energies using absolute rate theory.[184,185]

products with the opposite configuration.[337,338] Indeed, perfluorobenzene and benzene have opposite quadrupole moments.[339]

1.2.5   Unresolved problems

1.2.5.1   Reaction mechanism
The revised mechanisms shown in Schemes 1.45 and 1.46 for (pre)catalysts **1** and **2** were established through *reliable* computational chemistry,

in which full computational model and continuum solvent reaction field coupled with explicit solvation and extended basis sets whenever possible, were used. These computations were calibrated against available experimental results, including stoichiometric NMR experiments, X-Ray crystallographically characterized complexes (including off-loop species) and kinetics data (through Kozuch and Shaik's energetic span model approximation[203–208] based on TST[184,185] and some other approximations) in some cases.

It should be emphasized, however, that the accuracy of energetic aspects of many computational studies is ultimately both uncertain and unexamined. The potential error associated with free energy calculations likely increases with increasing size of the complex, the use of a less extended basis set and charge distribution, particularly in a continuum solvent. We note, however, that we do not believe this would alter our conclusions in any drastic way. Firstly, these energy-related aspects will not change the shape of the solution based PES, that is, the presence/absence of stationary and saddle points. In other words *multibond* concerted transformations represented by transition states **TS**$_a$ and **TS**$_c$ shown in Chart 1.4 have no mechanistic relevance, because they do not physically exist in solution as suggested by computations where implicit and/or explicit solvation is introduced. Secondly, we expect that the inaccuracy in free energy calculations is unlikely to exceed 20 kcal/mol and thus the H–H bond does not get cleaved via metal–ligand cooperation.[253]

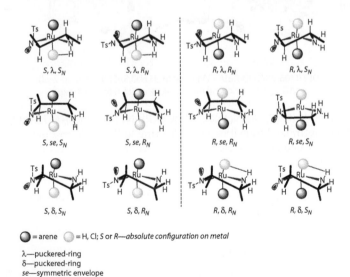

Chart 1.4 Isomers possible for the Noyori–Ikariya catalyst. (Adapted from Dub, P. A. et al., *Dalton Trans.*, 45, 6756–6781. Copyright 2016 Royal Society of Chemistry.)

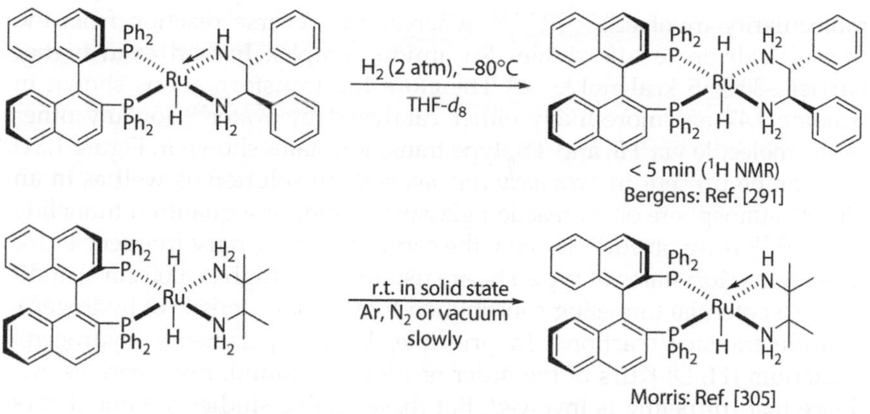

H_2 (2 atm), −80°C
THF-d_8

< 5 min (^{1}H NMR)
Bergens: Ref. [291]

r.t. in solid state
Ar, N_2 or vacuum
slowly

Morris: Ref. [305]

Scheme 1.47 Interpretation of chemical reactivity based on NMR experiment and/or visual observation. (Reprinted with permission from Dub, P. A. et al., *Dalton Trans.*, 45, 6756–6781. Copyright 2016 Royal Society of Chemistry.)

In 2006, Bergens' group reported that the 16e⁻ Ru amido complex [RuH{(R)-binap}{(R, R)-HN(CHPh)$_2$NH$_2$}] reacts with molecular hydrogen (~2 atm) at −80°C to afford *trans*-[RuH$_2${(R)-binap}{(R, R)-dpen}] as kinetic product in less than 5 min in THF-d_8 based on ^{1}H NMR experiment as shown in Scheme 1.47.[291] Similarly, Morris' group earlier reported in 2001 that yellow *trans*-[RuH$_2${(R)-binap}{tmen}] slowly loses H$_2$ under Ar, N$_2$ or vacuum to afford a dark-red 16e⁻ ruthenium hydridoamido complex [RuH{(R)-binap}{NHCMe$_2$CMe$_2$NH$_2$}] in the *solid state*, Scheme 1.7.[305] On the other hand, this complex reacted with molecular hydrogen (~1 atm) at room temperature to afford back the *trans*-dihydride in toluene-d_8.[305]

Can these experimental observations (assuming that everything is correctly assigned) be universally assumed to state that the H–H bond is cleaved/formed via metal–ligand cooperation? First, one must recognize that a catalytic reaction is not a stoichiometric reaction, and the so-called reaction path[198,340] of the former follows a MEP.[195–198] According to computational analysis for **1**, the MEP proceeds through the deprotonation of the η²-H$_2$ ligand by the RO⁻ anion (the anion of the not yet formed alcohol product or the conjugated anion of the solvent), an energetically feasible step in which the H–H bond is cleaved.[253] The first order saddle-point corresponding to cleavage of the H–H bond via metal–ligand cooperation is located extremely high on the PES, therefore this path is not realized during catalytic reaction.[253] Secondly, with respect to stoichiometric reaction, the computed relative *averaged* activation barrier of "direct" H–H cleavage via metal–ligand cooperation is also very high, roughly estimated to be ~19 kcal/mol (the calculated range is from 15 to 23 kcal/mol depending on the type of calculations, model used, and notably, the molecularity involved),[214,253,272,288] whereas the reverse reaction from the *trans*-dihydride to afford 16e⁻ Ru amido complex has an even higher barrier ~30 ± 5 kcal·mol⁻¹.[253,288] Therefore the transformations shown in Scheme 1.47 are more-likely either catalyzed by water[341] (or any other protic molecule via TS_f and TS_g type transition states shown in Figure 1.21) that can be present in *extremely tiny amounts* in solution as well as in an "inert" atmosphere or on reaction glassware, and/or a quantum tunneling effect[342–345] is involved.[346] Indeed, the computed imaginary frequencies for these transition states of type TS_b are usually very high (>1000 cm⁻¹), indicating a potential tunneling contribution that is characteristic of hydrogen/hydride transfer reactions. In principle, large experimental hydrogen/deuterium (H/D) KIE's of the order of 10–100, if found, may serve as evidence that tunneling is involved. But these kinetic studies are not always available, and if they are (e.g., Morris[272]), usually normal KIE's are found. Noyori and coworkers reported a combined KIE and kinetic solvent isotope effect (KSIE) of 2 under high KO-t-C_4H_9 concentration, but the reaction with H_2/$(CH_3)_2CHOH$ proceeds 50 times faster than that with D_2/$(CD_3)_2CDOD$ in the absence of base.[262] Under these conditions, tunneling maybe indeed involved in the *catalytic reaction*, but it essentially unclear in which step or steps. In fact, computations suggest that the neutral η^2-H_2 complex, an intermediate that precedes H–H bond cleavage via metal–ligand cooperation, is located 8–10 kcal/mol higher than transition states leading to catalyst regeneration via η^2-H_2 ligand deprotonation by RO⁻ anions.[253] Therefore even if tunneling is involved in H–H bond cleavage via metal–ligand cooperation, the catalytic reaction will not proceed through this path according to the computational analysis. What happens if the accuracy of energy calculations is now considered? Even if the numerical accuracy of DFT, the functional and basis set used, the continuum solvent model used, errors in enthalpy and entropy calculations, the software package used, the molecularity involved and other parameters exceeds 10 kcal·mol⁻¹, the H–H bond, assuming tunneling, won't be cleaved exclusively via metal–ligand cooperation. Instead H–H bond cleavage would occur most likely via a crossover of up to three paths originating from at least five possibilities to produce the reaction product from the inner-sphere ion-pair, obtained after the enantiodetermining hydride transfer step (EDS), Scheme 1.48.[253] The overall reaction can therefore be very complicated and the relative contribution of each pathway may vary with the extent of turnover, thus potentially leading to a change in the reaction rate.

1.2.5.2   *Enantioselectivity*
The Noyori–Ikariya (pre)catalyst **2** or similar complexes exist as mixtures of two diastereomers originating from the relative configuration on the metal atom.[347,348] For example, the real intermediate of the catalytic cycle, hydride complex RuH[(R, R)-N(Ts)CH(Ph)CH(Ph)NH₂](η⁶-p-cymene) exists

H–H cleavage and catalyst regeneration

Hydride transfer (EDS)

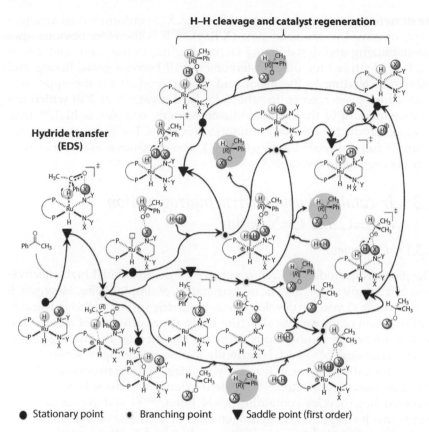

● Stationary point   ● Branching point   ▼ Saddle point (first order)

Scheme 1.48 "Reaction path"[198,340] (conceivable catalytic pathways and their crossover) of the catalytic hydrogenation of acetophenone with the active form of **1** using the assumption that tunneling is involved in so-called H–H bond cleavage via metal–ligand cooperation and that computational accuracy is >10 kcal/mol (X = Y = H and/or K). There are five possibilities to neutralize the anion of the product obtained after EDS hydride transfer and regenerate the catalyst via three associated TS's corresponding to H–H bond cleavage. Some selected stationary and saddle points are omitted for clarity. (Reprinted with permission from Dub, P. A. et al., *Dalton Trans.*, 45, 6756–6781. Copyright 2016 Royal Society of Chemistry.)

as a 99:1 mixture of S_{Ru} and R_{Ru}-diastereomers in toluene-d_8 at room temperature.[264] Taking into account the absolute configuration of the NTs nitrogen atom and three stable five-membered NN ring conformations,[349,350] there are 12 possible conformers available for the delivery of each *S*- and *R*-conFigured product, Chart 1.4.

The observed enantioselectivity with this catalyst thus likely originates from up to 12 enantiopathways combined from 24 transitions states. The statistical weight of each couple is unexplored and the main factors are drawn for possibly the most stable (S_{Ru}, λ, R_N)-conformational arrangement, observed in the solid-state (X-Ray).[264] It is therefore obvious, that the stabilizing and destabilizing factors leading to the major and minor product identified for this diastereomer, will become destabilizing and stabilizing leading to the major and minor product for the oppositely conFigured diastereomer. In other words, derivatives of **2** in which the relative amount of the opposite diastereomeric complex is high[351] may produce products with lower enantioselectivity. For some systems, for example fully alkylated analogues of **2**, the opposite diastereomer may become even more favorable.[352]

1.3 Ir-catalyzed asymmetric hydrogenation of C=C and C=N bonds

1.3.1 Overview

The pioneering studies of Crabtree et al.[353–355] demonstrated high effectiveness of Ir complexes in catalytic hydrogenation of simple olefins. However, it took another 15 years to develop first asymmetric variant of this reaction.[356] Optimization of the catalyst structure led to the appearance of reliable and widely applicable synthetic procedure for enantioselective hydrogenation of simple olefins[357–362] which is currently further extending its scope.[363–366]

Historically first Ir-catalyzed asymmetric hydrogenations used prochiral imines as substrates.[367,368] Evident extension of the substrates scope involved heterocycles containing C=N bond,[369–372] and recent achievements involve asymmetric hydrogenation of aromatic heterocycles,[373–376] various functionalized olefins,[377,378] as well as highly efficient asymmetric hydrogenation of ketones[379] and keto-esters[380,381]

1.3.2 Catalytic cycle and intermediates

In the early studies of nonenantioselective hydrogenation it has been found that *bis*-monophosphine Ir complexes give stable *trans*-solvate dihydrides **1** upon removal of the coordinated diene from a precatalyst. These solvate dihydrides were found to be capable of exchanging one or two of their solvent molecules for olefins yielding dihydride olefin complexes **2** or **3**, respectively which were characterized at $-80°C$, Equation 1.1.[354,355]

$$ \text{(1.1)} $$

$$ L = Ph_3P, \; \| = olefin, \; S = H_2O $$

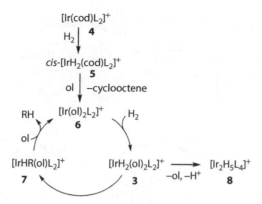

Figure 1.26 Catalytic cycle for Ir-catalyzed hydrogenation of olefins. (Data from Crabtree, R. H. et al., *J. Am. Chem. Soc.*, 104, 6994–7001, 1982.)

The catalytic cycle for simple olefins hydrogenation was also postulated in these early works of Osborn et al. It involves oxidative addition of H_2 to a precatalyst **4**, hydrogenation of the coordinated diene by the activated hydrogen in **5** that leads to a mixed cyclooctene–olefin complex **6**, fast oxidative addition of another molecule of H_2 yielding **3**, slow migratory insertion affording alkylhydrido complex **7**, and reductive elimination followed by coordination of olefin that recovers the catalyst–substrate complex **6** (Figure 1.26).

Binuclear iridium hydrides like **8** often form as off-loop species that can result in decreasing of the catalytic activity. Formation of trinuclear complexes completely deactivating the catalyst has been also observed.[382] Nevertheless, the dimeric hydrides formed reversibly before the elimination of the proton are catalytically active, since they can recover mononuclear dihydrides via reversible dissociation.[378,383]

Essentially the same catalytic cycle can operate in the case of Ir complexes of chiral chelating ligands. However, since in that case two donor atoms must be in *cis*-position, the intermediates in the catalytic cycle are less stable and must be characterized at low temperatures. Thus, Pfaltz and coworkers characterized dihydride **10** by low-temperature hydrogenation of [Ir(PHOX)(cod)]$^+$BAr$_F^-$ **9** (Scheme 1.49).[384] Raising the temperature resulted in the replacement of the diene with two molecules of donating solvent yielding two diastereomers of the solvate dihydride **11** and **12**. Totally four isomers of **10** and four isomers of **11** are possible, hence this transformation is diastereoselective due to a strong electronic influence of the coordinating N and P atoms, favoring addition of a hydride *trans* to the Ir-N bond.

The same research group recently characterized key intermediates in the Ir-catalyzed asymmetric hydrogenation of simple olefins.[385] When a

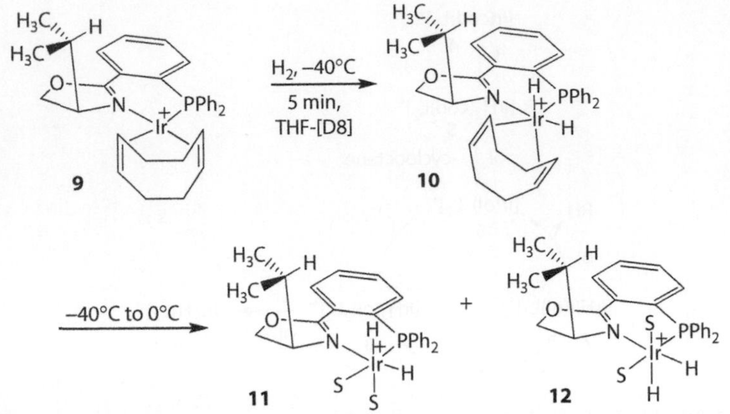

Scheme 1.49 Low-temperature hydrogenation of [Ir(PHOX)(cod)]⁺BArF⁻ **9**. (From Mazet, C. et al., *J. Am. Chem. Soc.*, 126, 14176–14181, 2004. With permission.)

CD₂Cl₂ solution of the iridium complex **13** and excess of partially deuterated olefin **14** was treated with three bar of hydrogen gas at 233 K for 35 min and then degassed, 2D NMR two isomeric dihydride alkene iridium complexes **15a, b** in a ratio of about 11:1 (Scheme 1.50).[385]

At −20°C **15a** and **15b** demonstrated rapid exchange with each other and the free alkene **14-[D₅]**, thus indicating fast equilibration between isomers via alkene dissociation/association. This evidently means that other less stable isomers may be kinetically accessible.

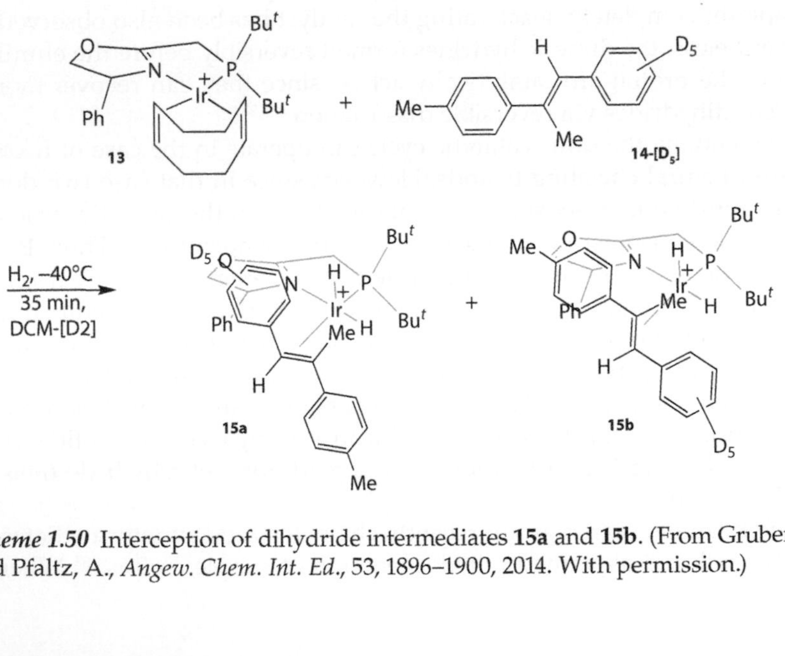

Scheme 1.50 Interception of dihydride intermediates **15a** and **15b**. (From Gruber, S. and Pfaltz, A., *Angew. Chem. Int. Ed.*, 53, 1896–1900, 2014. With permission.)

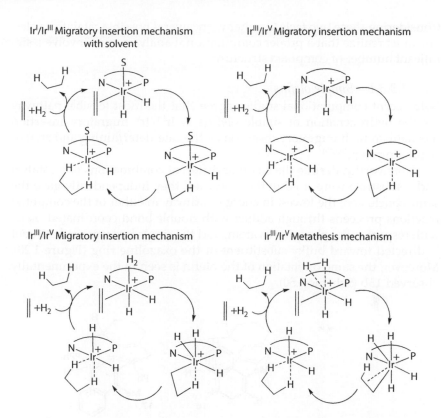

Figure 1.27 Four possible mechanisms of the Ir-catalyzed hydrogenation of simple olefins.

To convert an intermediate similar to **15a, b** into the reaction product, the presence of additional hydrogen gas was necessary.[385] This was interpreted as an experimental evidence for the so-called Ir^{III}/Ir^V route via an $[Ir^{III}(H)_2(alkene)(H_2)(L)]^+$ intermediate, as originally proposed by groups of Andersson and Burgess.[386–389] Since the results of experimental mass-spectral investigation demonstrated also possibility of a Ir^I/Ir^{III} route that does not involve participation of an additional H_2 molecule, the complete mechanistic investigation must consider competition of several catalytic cycles (Figure 1.27).

1.3.3 Mechanism of enantioselection

The most important feature of the Ir-catalyzed asymmetric hydrogenation is the impossibility of a chelate binding for the vast majority of the substrates. Besides, the most effective catalysts in this field have C_1-symmetry.

Considering these two facts together with several possible catalytic cycles, one must realize that a proper computational analysis must involve a significant number of computed structures.

1.3.3.1   Simple olefins

Both recent computational studies agree that the most feasible pathway for the hydrogenation of simple olefins is Ir^{III}/Ir^V migratory insertion mechanism with migratory insertion as the rate-determining and stereo-regulating step.[390,391]

Noteworthy, despite the fact that different combinations of a catalyst and a substrate were used in these studies, they independently gave the same conclusion: the lowest in energy pathway in either of the computed reactions proceeds through adduct with double bond coordinated *trans*-with respect to the phosphorus atom, and the single hydrogen substituent is directed toward bulky substituent of the oxazoline ring (Figure 1.28). Moreover, the same orientation of the olefin is seen in the experimentally observed **15b** (Scheme 1.50).

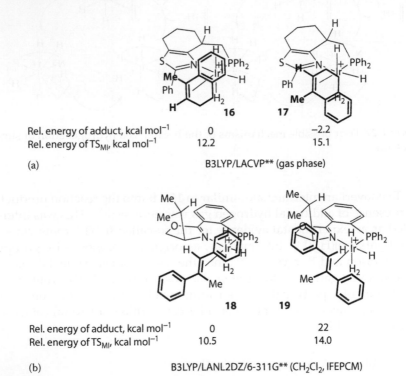

Rel. energy of adduct, kcal mol^{-1}	0	−2.2
Rel. energy of TS$_{MI}$, kcal mol^{-1}	12.2	15.1

(a)   B3LYP/LACVP** (gas phase)

Rel. energy of adduct, kcal mol^{-1}	0	22
Rel. energy of TS$_{MI}$, kcal mol^{-1}	10.5	14.0

(b)   B3LYP/LANL2DZ/6-311G** (CH$_2$Cl$_2$, IFEPCM)

Figure 1.28 Structures of the $[Ir(H)_2(olefin)(H_2)]^+$ complexes and relative energies for them and corresponding transition states of migratory insertion computed in (a)[390] and (b).[391] The 3D structures of these transition states are available on the CD.

The Andersson's group also argued that since the Ir-catalyzed hydro-genation is most selective for trisubstituted olefins, the placement of the hydrogen substituent may play a crucial role in stereo-discrimination. This suggestion resulted in a qualitative rule for the prediction of the sense of enantioselection in the asymmetric hydrogenation of a nonche-lating trisubstituted olefin by a chiral [(N∩P)Ir(COD)]+ cation.[390]

The application of this rule is most straightforward in the case of the ligands like PHOX, with chirality on sp³-carbon atom. In that case the ori-entation of the stereo-regulating substituent is evident, and to predict the sense of enantioselection of a trisubstituted olefin, one should just draw the olefin with the hydrogen atom pointed down-left as in **15b** or up-left as in **18** (Figure 1.29).

If the ligand has a backbone chirality like in **16**, computations are required to determine, whether the stereo-discriminating substituent on the oxazoline ring lies "below" or "above the ligand plane.[390]

This approach has been proved applicable for other types of ligands and trisubstituted olefins containing various functional groups as one of the substituents[366,392] including cyclic sulfones.[365]

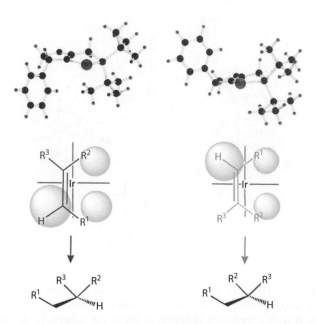

Figure 1.29 Empirical rule for elucidating handedness of the hydrogenation product in the Ir-catalyzed asymmetric hydrogenation of trisubstituted olefins. The most facile migratory insertion takes place if the olefin is coordinated *trans*-to phosphorus.

1.3.3.2 Imines

In the initial mechanistic studies of the Ir-catalyzed hydrogenation of imines catalytic cycles involving imine coordination to the metallic center have been considered.[393,394] However, in the recent computational study of Hopmann and Bayer it has been shown that in the case of asymmetric hydrogenation with [Ir(PHOX)(cod)]$^+$BAr$_F$$^-$ 9 various catalytic cycles containing coordinated imine are characterized with unreasonably high activation barriers (35–50 kcal/mol). On the other hand, a catalytic cycle resembling that outlined in 2006 by Oro et al. (Figure 1.30)[395,396] has been found kinetically feasible.

The rate-limiting and stereo-determining stage in this catalytic cycle is the hydride transfer to the iminium salt. Since the iminium salt in the intermediate **C** is not bound to the iridium atom, additional orientations of the C=N double bond are available. Computational study showed that the enantioselectivity is determined in the competition of transition states in either of which the NH-hydrogen is positioned nearby the largest substituent (Figure 1.31). The reasons for the different stabilities of iminium cations and corresponding transition states are far from evident in this case. Hence, computations of similar level are necessary for each particular combination of a catalyst and a substrate.

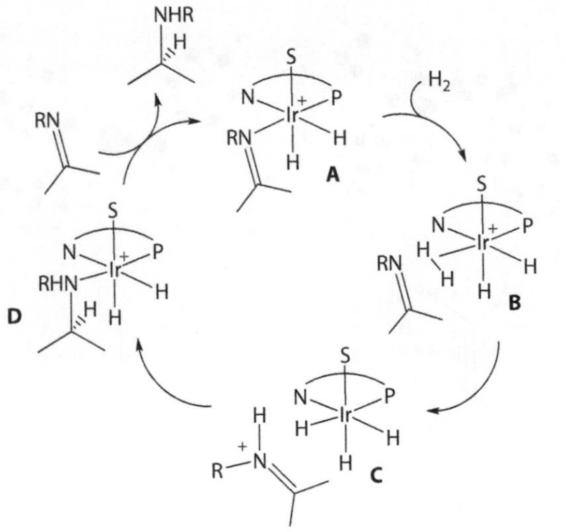

Figure 1.30 Catalytic cycle for the Ir-catalyzed asymmetric hydrogenation of imines. Dihydrogen replaces coordinated imine in **A** yielding molecular hydrogen complex **B** with a molecule of imine in the outer sphere. Then imine is protonated by one of the hydrogen atoms from the H$_2$ molecule to give neutral Ir^{III} complex **C**. The following hydride transfer of the hydride *trans-* to phosphorus gives the catalyst–product complex **D**. Replacement of the product with a new substrate molecule finishes the catalytic cycle recovering **A**.

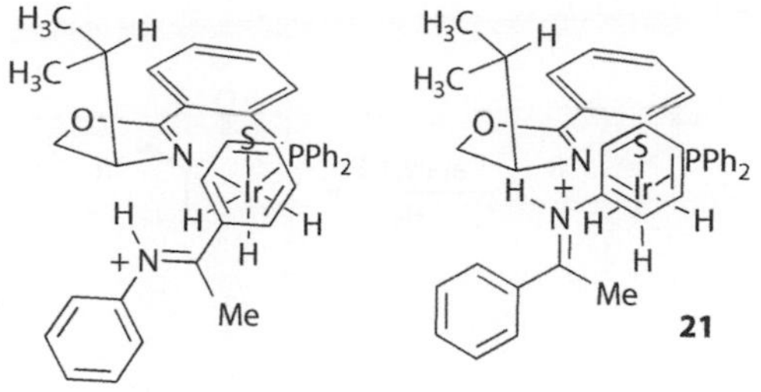

| Rel. free energy of adduct, kcal mol^{-1} | 0 | |
| Rel. free energy of TS$_{MI}$, kcal mol^{-1} | 10.4 | 12.1 |

B3LYP/LANL2DZ/6-311G** (CH$_2$Cl$_2$, IFEPCM)

Figure 1.31 Iminium cations **20, 21** preceding two competing transition states and their relative free energies.[391] The 3D structures of these transition states are available on the CD.

1.3.3.3   Heterocycles

Although recently discovered Ir-catalyzed asymmetric hydrogenations of isoquinolinium[374–376] and pyridinium[375] salts are undoubtedly very interesting form the mechanistic point of view, even the details of the catalytic cycle are yet unknown and there are controversies on this point in the literature.

On the other hand, the mechanism of the asymmetric hydrogenation of cyclic lactones and lactames containing exocyclic double bond (Scheme 1.51) has been recently studied.[378]

NMR Studies showed that removal of cod ligand leads to formation of a dimer existing in dynamic equilibrium with monomeric solvate dihydrides. Computational study has been carried out assuming monomeric catalyst (since the substrate could not be accommodated on the dimer) and Ir^I/Ir^{III} mechanism (since stable molecular H$_2$ complexes were not located).

Computational study considered only pathways in which the migratory insertion takes place in a complex with the double bond in equatorial position oriented coplanar to the P–Ir–H unit. Hydride transfer to the protonated carbon atom of the double bond has been computed to be relatively facile. Therefore only competition of the migratory insertion in adducts **23–26** was considered to contribute into the observed ee values. (Scheme 1.52).

The energy difference between the lowest in energy migratory insertion TS formed through the adduct **23** and TS originating from the adduct **25** leading to the opposite enantiomer was computed to be 2.1 kcal/mol that nicely corresponds to the experimentally observed ee value. The substrate dependence of the optical yields was also reasonably reproduced (Scheme 1.51).

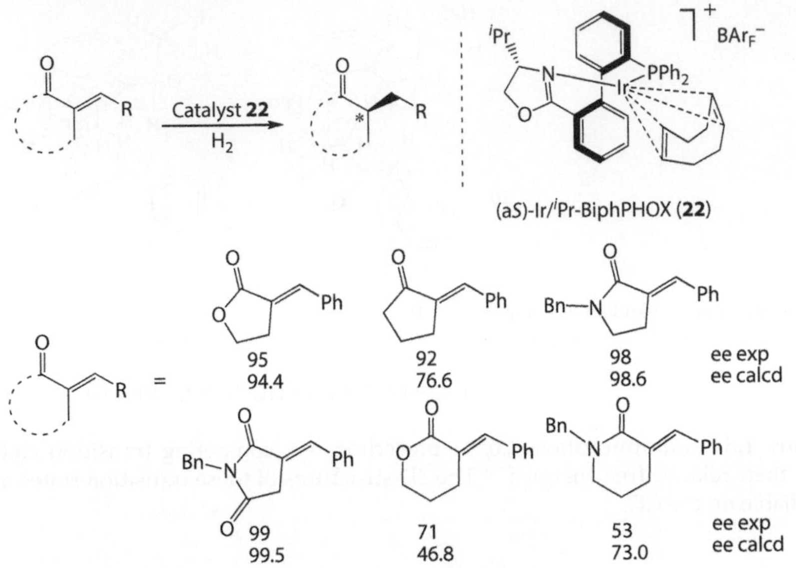

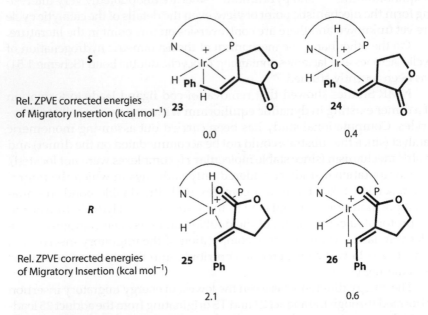

Scheme 1.51 Asymmetric hydrogenation of exocyclic α,β-unsaturated carbonyl compounds with an Ir/BiphPhox catalyst.

Scheme 1.52 Substrate coordination in four competing pathways in the asymmetric hydrogenation of exocyclic α,β-unsaturated carbonyl compounds with an Ir/BiphPhox catalyst.

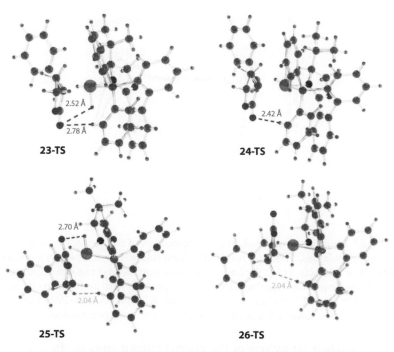

23-TS   **24-TS**

25-TS   **26-TS**

Figure 1.32 Optimized structures of the migratory insertion transition states in the asymmetric hydrogenation of exocyclic α,β-unsaturated carbonyl compounds with an Ir/BiphPhox catalyst. The 3D structures of these transition states are available on CD.

In both **23** and **25** (and corresponding transition states) hydrogen bonding between the axial hydride and the oxygen of the carbonyl group in the substrate is possible (Figure 1.32). Additionally in the **23-TS** a stabilizing nonclassical C-H...O bonding is observed. The same interaction is impossible in **25-TS**. Instead a close contact of the same aromatic proton with CH_2 group of the heterocycle makes **25-TS** less stable (Figure 1.32).

On the other hand, **26-TS** leading to *R*-product was computed to be very close in stability to the lowest in energy **23-TS** or **24-TS** (Scheme 1.52). Hence, the experimentally observed *S*-enantioselective reaction means that the migratory insertion via the **26-TS** does not actually take place.

In search of a possible reason for this, the approach of the substrate to the catalyst resulting in the coordination of the substrate appropriate for the occurrence of the migratory insertion via **23-TS** and **26-TS** was simulated. The corresponding energy scans are shown in Figure 1.33.

As one can see in Figure 1.33, when approaching to **23**, the substrate rapidly finds a minimum via hydrogen bond formation with the axial hydride. On the other hand, in the approach to **26** such interaction is

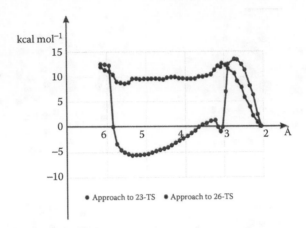

Figure 1.33 Energy scans simulating the approach of the substrate to the corresponding Ir hydrides. The final minima correspond to the completely optimized structures of adducts. (Liu, Y. et al.: Mechanism of the asymmetric hydrogenation of exocyclic α,β-unsaturated carbonyl compounds with an iridium/BiphPhox catalyst: NMR and DFT studies. *Angew. Chem. Int. Ed.* 2014. 53, 1901–1905. Copyright Wiley-VCH Verlag GmbH & Co. KGaA. Reproduced with permission.)

reliably switched off by one of the phenyl substituents of the catalyst.[364] As a result, only **25-TS** is competing with **23-TS** and **24-TS**.

1.3.4   Conclusion

The origins of enantioselection in the Ir-catalyzed asymmetric hydrogenation of simple olefins are now reasonably well understood as a result of intensive experimental and computational studies. This helps in development of new effective catalysts and synthetic methods. Similar progress is wanted for the recently discovered highly effective Ir-catalyzed asymmetric hydrogenations of aromatic heterocycles, ketones, and keto-esters.

1.4   Pd-catalyzed asymmetric hydrogenation of indoles

1.4.1   Overview

Chiral indolines represent useful building blocks in pharmaceuticals, herbicides and insecticides.[397] Asymmetric hydrogenation of unprotected indoles with molecular hydrogen represents the most straightforward and atom-economy method to produce these compounds.[398,399] Although the asymmetric (up to 98% ee) hydrogenation of N-protected or N–Ts indoles with Rh,[400–402] and Ru[403] is known since 2000 (R. Kuwano, Kyushu University) or Ir[404] since 2010 (A. Pfaltz, University of Basel), the very first example of hydrogenation of unprotected indoles was discovered in X. Zhang and Y.-G. Zhou groups in 2010.[405] The catalyst system composed of Pd(OCOCF$_3$)$_2$/chiral-substituted BINAP in the combination with Bronsted acid (1 equiv relative to the substrate) affords enantioenriched indolines with up to 96%–98% ee.[406] The better conversion and enantioselectivity is achieved in the combined TFE/DCM solvent (TFE = 1,2,3-trifluoroethanol, DCM = dichloromethane) suggesting that both hydrogen-bonding and low-polarity are important factors in this reaction with respect to conversion and enantioselectivity. The relative drawback of the method is high substrate-to-catalyst ratio (S/C) = 50, however, the system is tolerant to oxygen, acid and water.

1.4.2 Catalytic cycle

The proposed catalytic cycle is shown in Figure 1.34.[406] The catalytic cycle starts with intermediate **a**, obtained from [Pd(OCOCF$_3$)$_2${(R)-BINAP}] via dissociation of one ⁻OCOCF$_3$. The reaction of intermediate **a** with molecular H$_2$ (possibly solvent-assisted) affords monohydride intermediate **b**. Substrate (2-methylindole in Figure 1.34) enters the cycle in the protonated form (from organic acid presented at stoichiometric amount) of iminium cation α, which forms hydrogen-bonded adduct **c**. After rate- and enantio-determining hydride transfer, catalyst-product complex **d** is formed. The cycle restarts after product dissociation into solution.

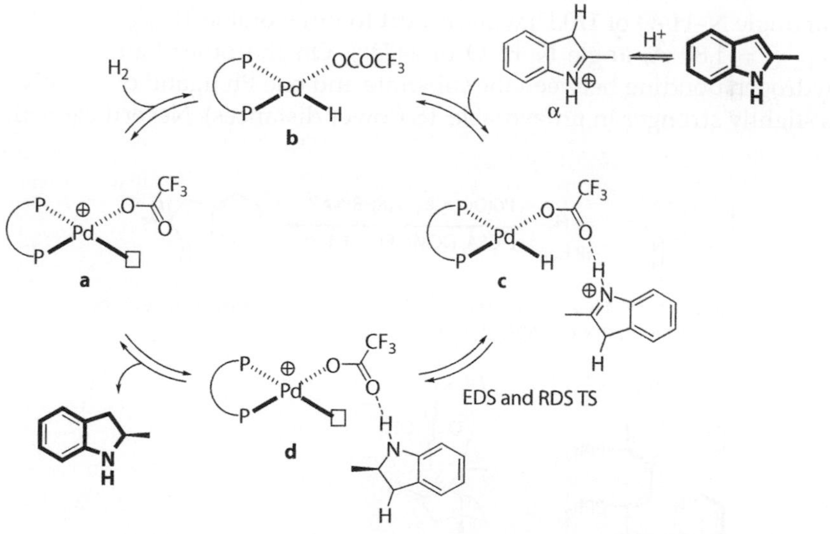

Figure 1.34 Proposed mechanism for the asymmetric hydrogenation of 2-methylindole.

The overall mechanism thus is outer-sphere and ionic consisting of H$^+$ substrate protonation from acid and then H$^-$ transfer from the catalyst (H$^+$/H$^-$ sequence).

1.4.3   Enantioselectivity

Full DFT/B3LYP/LACVP(Pd)/6-31G**(C, H, O, N, F, P) gas-phase model has been computationally considered[406] in order to understand the factors governing the enantioselectivity of 2-methylindole hydrogenation into (R)-2-methylindoline. Experimental details,[406] determined sense of enantioselectivity is shown in Scheme 1.53. Chiral L-camphorsulfonic acid (L-CSA) does not seem to play an important role in the enantiodetermining step, since similar ee is obtained in the presence of TsOH·H$_2$O, for example.

According to the gas-phase computational model, hydride transfer represents both rate-determining and enantio-determining step that proceeds via eight-membered-ring transition state structure. Two diastereomeric transition states leading to major (R)-2-methylindoline and minor (S)-2-methylindoline are compared in Figure 1.35. Both transition states are stabilized by relatively strong N–H···O=C(OPd)CF$_3$ hydrogen bonding and apparently weaker C–H...π hydrogen bonding between the substrate and one Ph ligand of the BINAP (the distance between H(CH$_2$) and center of Ph is 2.85 Å (favorable TS) and 2.62 Å (favorable TS), respectively). The strong N–H···O=C(OPd)CF$_3$ hydrogen bonding interaction seems to be *stronger* in the favorable TS ($d_{N...O}$ = 2.65 Å, $d_{NH...O}$ = 1.60 Å, almost linear angle N–H–O of 170.1°) with respect to unfavorable TS ($d_{N...O}$ = 2.72 Å, $d_{NH...O}$ = 1.80 Å, angle N–H–O of 147.1°). On the other hand, C–H...π hydrogen bonding between the substrate and one Ph ligand of the BINAP is slightly stronger in unfavorable TS (lower distances). Nevertheless, this

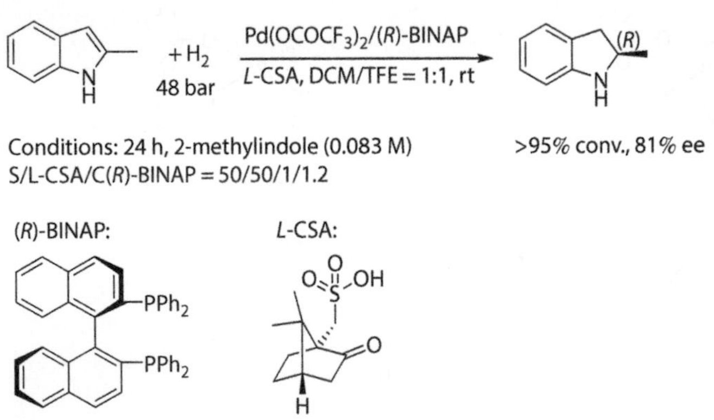

Conditions: 24 h, 2-methylindole (0.083 M)   >95% conv., 81% ee
S/L-CSA/C(R)-BINAP = 50/50/1/1.2

Scheme 1.53 Catalytic hydrogenation of 2-Methylindole

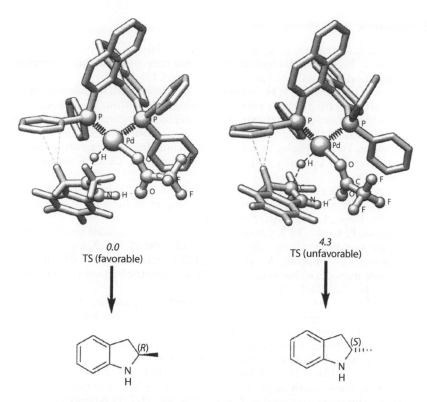

0.0
TS (favorable)

4.3
TS (unfavorable)

Figure 1.35 [PdH(OCOCF$_3$){(R)-BINAP}]–2-methylindole: two diastereomeric transition states leading to major (R)-2-methylindoline and minor (S)-2-methylindoline. $\Delta G(solv)_{298\ K°}$ = E(SCF) + ZPE – TS(*gas*) + G^{solv}, kcal/mol. Selected hydrogen atoms are omitted for clarity.

effect does not compensate much stronger N–H...O=C(OPd)CF$_3$ hydrogen bonding interaction which is thus the major factor responsible for the enantioselectivity. The calculated energy difference ($\Delta G(solv)_{298\ K°}$) between diastereomeric TS's of 4.3 kcal/mol corresponds to calculated enantioselectivity of 99.9% ee ($RT = 0.58$ kcal/mol) and reproduces the sense of enantioselection. The term $G(solv)_{298\ K°}$ has been calculated as combination of gas-phase components plus single-point energy with continuum model: E(SCF) + ZPE – TS(*gas*) + G^{solv}. Most-likely, additional solvation of these TS' via nonspecific and specific solvation contributes to the experimental ee of 80% (experimental studies suggest that solvent choice is important to achieve higher ee's). This could change the relative strength of H...O=C(OPd)CF$_3$ hydrogen bonding in each diastereomeric TS' (not necessarily with the same degree and extension) and lower the experimental ee.

References

1. Young, J. F.; Osborn, J. A.; Jardine, F. H.; Wilkinson, G. Hydride interme- diates in homogeneous hydrogenation reactions of olefins and acetylenes using rhodium catalysts. *Chem. Commun.* 1965, 131–132.
2. Osborn, J. A.; Jardine, F. H.; Young, J. F.; Wilkinson, G. The preparation and properties of tris(triphenylphosphine)halogeno-rhodium(I) and some reac- tions thereof including catalytic homogeneous hydrogenation of olefins and acetylenes and their derivatives. *J. Chem. Soc. A* 1966, 1711–1732.
3. Knowles, W. S.; Sabacky, M. J. Catalytic asymmetric hydrogenation employ- ing a soluble, optically active, rhodium complex. *Chem. Commun.* 1968, 1445–1446.
4. Horner, L.; Siegel, H.; Büthe, H. Asymmetric catalytic hydrogenation with an optically active phosphinerhodium complex in homogeneous solution. *Angew. Chem., Int. Ed. Engl.* 1968, 7, 942–943.
5. Lagasse, F.; Kagan, H. B. Chiral monophosphines as ligands for asymmetric organometallic catalysis. *Chem. Pharm. Bull.* 2000, 48, 315–324.
6. Komarov, I. V.; Borner, A. Highly enantioselective or not?—Chiral mono- dentate monophosphorus ligands in the asymmetric hydrogenation. *Angew. Chem. Int. Edit.* 2001, 40, 1197–1200.
7. Blaser, H. U.; Malan, C.; Pugin, B.; Spindler, F.; Steiner, H.; Studer, M. Selective hydrogenation for fine chemicals: Recent trends and new develop- ments. *Adv. Synth. Catal.* 2003, 345, 103–151.
8. Tang, W.; Zhang, X. New chiral phosphorus ligands for enantioselective hydrogenation. *Chem. Rev.* 2003, 103, 3029–3069.
9. Halpern, J.; Riley, D. P.; Chan, A. S. C.; Pluth, J. J. Novel coordination chemistry and catalytic properties of cationic 1,2-bis(diphenylphosphino)- ethanerhodium(I) complexes. *J. Am. Chem. Soc.* 1977, 99, 8055–8057.
10. Brown, J. M.; Chaloner, P. A. Mechanism of asymmetric hydrogenation catalysed by rhodium(I) DIOP complexes. *J. Chem. Soc., Chem. Commun.* 1978, 321–322.
11. Landis, C. R.; Halpern, J. Asymmetric hydrogenation of methyl (Z)-α- acet- amidocinnamate catalyzed by {1,2-bis((phenyl-*o*-anisyl)phosphino) ethane} rhodium(I): Kinetics, mechanism, and origin of enantioselection. *J. Am. Chem. Soc.* 1987, 109, 1746–1754.
12. Gridnev, I. D.; Imamoto, T.; Hoge, G.; Kouchi M.; Takahashi, H. Asymmetric hydrogenation catalyzed by a rhodium complex of (R)-(*tert*- butylmethylphosphino)-(di-*tert*-butylphosphino)methane: Scope of enanti- oselectivity and mechanistic study. *J. Am. Chem. Soc.* 2008, 130, 2560–2572.
13. Brown, J. M.; Chaloner P. A.; Morris, G. A. Identification of a further tran- sient species relating to rhodium-complex catalysed asymmetric hydroge- nation. *J. Chem. Soc., Chem. Commun.* 1983, 664–666.
14. Brown, J. M.; Chaloner, P. A. The catalytic resting state of asymmetric homogeneous hydrogenation. Exchange processes delineated by nuclear magnetic resonance saturation-transfer (DANTE) techniques. *J. Chem. Soc., Perkin Trans. II* 1987, 1583–1588.
15. Ramsden, J. A.; Claridge, T. D. W.; Brown, J. M. Structure and dynamics of intermediates in asymmetric hydrogenation by rhodium complexes of (2-methoxyphenyl)-P-phenyl-P-(2′-diphenylphosphino)ethylphosphine. *J. Chem. Soc., Chem. Commun.* 1995, 2469–2471.

16. Imamoto, T.; Tamura, K.; Zhang, Z.; Horiuchi, Y.; Sugiya, M.; Yoshida, K.; Yanagisawa, A.; Gridnev, I. D. Rigid P-chiral phosphine ligands with *tert*-butyl groups for rhodium-catalyzed asymmetric hydrogenation of functionalized alkenes. *J. Am. Chem. Soc.* 2012, *134*, 1754–1769.

17. Gridnev, I. D. Recent experimental and computational studies of the mechanisms of enantioselection in asymmetric catalytic reactions T. *J. Synth. Org. Chem. Jpn.* 2014, *74*, 1250–1264.

18. Gridnev, I. D.; Imamoto, T. Challenging the major/minor concept in Rh-catalyzed asymmetric hydrogenation. *ACS Catal.* 2015, *5*, 2911–2915.

19. Bircher, H.; Bender, B. R.; von Philipsborn, W. Interconversion of diastereomeric complexes involved in Rh-catalysed asymmetric hydrogenation: A EXSY NMR study. *Magn. Res. Chem.* 1993, *31*, 293–298.

20. Gridnev, I. D.; Liu, Y.; Imamoto, T. Mechanism of asymmetric hydrogenation of β-dehydroamino acids catalyzed by rhodium complexes: Large-scale experimental and computational study. *ACS Catal.* 2014, *4*, 203–219.

21. Gridnev, I. D.; Imamoto, T. *Izv. Akad. Nauk, Ser. Khim.* 2016, 1514–1534.

22. Gridnev, I. D.; Kohrt, C.; Liu, Y. Direct experimental and computational evidence for the dihydride pathway in TangPHOS-Rh catalysed asymmetric hydrogenation. *Dalton Trans.* 2014, *43*, 1785–1790.

23. Gridnev, I. D.; Higashi, N.; Asakura, K.; Imamoto, T. Mechanism of asymmetric hydrogenation catalyzed by a rhodium complex of (*S, S*)-bis(*t*-butylmethylphosphino)ethane. Dihydride mechanism of asymmetric hydrogenation. *J. Am. Chem. Soc.* 2000, *122*, 7183–7194.

24. Gridnev, I. D.; Yamanoi, Y.; Higashi, N.; Tsuruta, H.; Yasutake, M.; Imamoto, T. Asymmetric hydrogenation catalyzed by (*S, S*)-BisP*-Rh and (*R, R*)-MINIPHOS-Rh complexes: Scope, limitations and mechanism. *Adv. Synth. Catal.* 2001, *343*, 118–136.

25. Giernoth, R.; Heinrich, H.; Adams, N. J.; Deeth, R. J.; Bargon, J.; Brown, J. M. PHIP detection of a transient rhodium dihydride intermediate in the homogeneous hydrogenation of dehydroamino acids. *J. Am. Chem. Soc.* 2000, *122*, 12381–12382.

26. Heinrich, H.; Giernoth, R.; Bargon, J.; Brown, J. M. Observation of a stable *cis*-diphosphine solvate rhodium dihydride derived from PHANEPHOS. *Chem. Commun.* 2001, 1296–1297.

27. Gridnev, I. D.; Higashi, N.; Imamoto, T. Formation of a stable rhodium(III) dihydride complex and its reactions with prochiral substrates of asymmetric hydrogenation. *Organometallics* 2001, *20*, 4542–4553.

28. Imamoto, T.; Yashio, K.; Crepy, K. V. L.; Katagiri, K.; Takahashi, H.; Kouchi, M.; Gridnev I. D., P-chiral tetraphosphine dirhodium complex as a catalyst for asymmetric hydrogenation: Synthesis, structure, enantioselectivity, and mechanism. stereoselective formation of a dirhodium tetrahydride complex and its reaction with methyl (Z)-α-acetamidocinnamate. *Organometallics* 2006, *25*, 908–914.

29. Halpern, J. Mechanism and stereoselectivity of asymmetric hydrogenation. *Science* 1982, *217*, 401–407.

30. Brown, J. M. *Hydrogenation of Functionalized Carbon-Carbon Double Bonds*; E. N. Jacobsen, A. Pfaltz and H. Yamamoto, Eds.; Springer: Berlin, Germany, 1999; Vol. 1, pp. 119–182.

31. Landis, C. R.; Hilfenhaus, P.; Feldgus, S. Structures and reaction pathways to rhodium(I)-catalyzed hydrogenation of enamides. A model DFT study. *J. Am. Chem. Soc.* 1999, *121*, 8741–8754.

32. Feldgus, S.; Landis, C. R. Large-scale computational modeling of [Rh(DuPHOS)]-catalyzed hydrogenation of prochiral enamides: Reaction pathways and the origin of enantioselection. *J. Am. Chem. Soc.* 2000, *122*, 12714–12727.

33. Feldgus, S.; Landis, C. R. Origin of enantioreversal in the rhodium-catalyzed asymmetric hydrogenation of prochiral enamides and the effect of α-substituent. *Organometallics* 2001, *20*, 2374–2386.

34. Landis, C. R.; Feldgus, S. A simple model for the origin of enantioselection and the anti "lock-and-key" motif in asymmetric hydrogenation of enamides as catalyzed by chiral diphosphine complexes of Rh(I). *Angew. Chem. Int. Ed.* 2000, *39*, 2863–2866.

35. Brown, J. M.; Chaloner, P. A. Structural characterisation of a transient intermediate in rhodium-catalysed asymmetric homogeneous hydrogenation. *J. Chem. Soc. Chem. Commun.* 1980, 344–346.

36. Brown, J. M.; Chaloner, P. A.; Parker, D. Catalytic aspects of metal phosphine complexes. In *Advances in Chemistry Series*, Vol. *196*, Alyea, E. C.; Meek, D. W., Eds.; American Chemical Society, Washington, DC, 1982, pp. 355–369.

37. Gridnev, I. D.; Higashi, N.; Imamoto, T. On the origin of opposite stereoselection in the asymmetric hydrogenation of phenyl- and *tert*-butyl-substituted enamides. *J. Am. Chem. Soc.* 2000, *122*, 10486–10487.

38. Gridnev, I. D.; Yasutake, M.; Higashi, N.; Imamoto, T. Asymmetric hydrogenation of enamides with Rh-BisP* and Rh-MINIPHOS catalysts. Scope, limitations, and mechanism. *J. Am. Chem. Soc.* 2001, *123*, 5268–5276.

39. Yasutake, M.; Gridnev, 1.I. D.; Higashi, N.; Imamoto, T. Highly enantioselective hydrogenation of (*E*)-β-(acylamino)acrylates catalyzed by Rh(I)-complexes of electron-rich P-chirogenic diphosphines. *Org. Lett.* 2001, *3*, 1701–1704.

40. Gridnev, I. D.; Higashi, N.; Imamoto, T. Interconversion of monohydride intermediates in Rh(I)-catalyzed asymmetric hydrogenation of dimethyl 1-benzoyloxyethenephosphonate. *J. Am. Chem. Soc.* 2001, *123*, 4631–4632.

41. Gridnev, I. D.; Yasutake, M.; Imamoto, T.; Beletskaya, I. P. Asymmetric hydrogenation of α,β-unsaturated phosphonates with Rh-BisP* and Rh-MiniPHOS catalysts: Scope and mechanism of the reaction. *Proc. Natl. Acad. Sci. U.S.A.* 2004, *101*, 5385–5390.

42. Gridnev, I. D.; Imamoto, T. On the mechanism of stereoselection in Rh-catalyzed Asymmetric hydrogenation: A general approach for predicting the sense of enantioselectivity. *Acc. Chem. Res.* 2004, *37*, 633–644.

43. Gridnev, I. D.; Imamoto, T. Mechanism of enantioselection in Rh-catalyzed asymmetric hydrogenation. The origin of utmost catalytic performance. *Chem. Commun.* 2009, 7447–7464.

44. Imamoto, T.; Itoh, T.; Yoshida K.; Gridnev, I. D. Marked isotopic effects of the optical yields in rh-catalyzed asymmetric hydrogenation of enamides *Chem. Asian J.* 2008, *3*, 1636–1641.

45. Seebach, D.; Plattner, D. A.; Beck, A.; Wang, Y. M.; Hunziker, D.; Petter, W. On the mechanisms of enantioselective reactions using α,α,α'α'-tetraaryl-1,3-dioxane-4,5-dimethanol (TADDOL)-derived titanates: Differences between C_2- and C_1-symmetrical TADDOLs – facts, implications and generalizations. *Helv. Chim. Acta* 1992, *75*, 2171–2209.

46. Koenig, K. E.; Sabacky, M. J.; Bachman, G. L.; Christopfel, W. C.; Barnstorff, H. D.; Friedman, R. B.; Knowels, W. S.; Stults, B. R.; Vineyard, B. D.; Weinkauff, D. J. Asymmetric hydrogenations with rhodium chiral phosphine catalysts. *Ann. N Y Acad. Sci.* 1980, *333*, 16–22.
47. Knowels, W. S. Asymmetric hydrogenation. *Acc. Chem. Res.* 1983, *16*, 106–112.
48. Knowles, W. S. Asymmetric hydrogenations (Nobel lecture). *Angew. Chem. Int. Ed.* 2002, *41*, 1998–2007.
49. Wada, Y.; Imamoto, T.; Tsuruta, H.; Yamaguchi, K.; Gridnev, I. D. Optically pure 1,2-bis[(o-alkylphenyl)phenylphosphino]ethanes and their use in rhodium-catalyzed asymmetric hydrogenations of alpha-(acylamino)acrylic derivatives. *Adv. Synth. Catal.* 2004, *346*, 777–788.
50. Tsuruta, H.; Imamoto, T.; Yamaguchi, K.; Gridnev, I. D. Evidence for the importance of conformational equilibria in Rh-diphosphine complexes for the enantioselection in Rh-catalyzed asymmetric hydrogenation. *Tetrahedron Lett.* 2005, *46*, 2879–2882.
51. Vineyard, B. D.; Knowels, W. S.; Sabacky, G. L.; Bachman, G. L.; Weinkauff, D. J. Asymmetric hydrogenation. Rhodium chiral biphosphine catalyst. *J. Am. Chem. Soc.* 1977, *99*, 5946–5952.
52. Imamoto, T.; Sugita, K.; Yoshida, K. An air-stable P-chiral phosphine ligand for highly enantioselective transition-metal-catalyzed reactions. *J. Am. Chem. Soc.* 2005, *127*, 11934–11935.
53. Miyashita, A.; Yasuda, A.; Takaya, H.; Toriumi, K.; Ito, T.; Souchi T.; Noyori, R. Synthesis of 2,2'-bis(diphenylphosphino)-1,1'-binaphthyl (BINAP), an atropisomeric chiral bis(triaryl)phosphine, and its use in the rhodium(I)-catalyzed asymmetric hydrogenation of alpha-(acylamino)acrylic acids. *J. Am. Chem. Soc.* 1980, *102*, 7932–7934.
54. Takaya, T.; Ogasawara, M.; Hayashi, T.; Sakai, M.; Miyaura, N. Rhodium-catalyzed asymmetric 1,4-addition of aryl- and alkenylboronic acids to enones. *J. Am. Chem. Soc.* 1998, *120*, 5579–5580.
55. de Vries, J. G.; Lefort, L. The combinatorial approach to asymmetric hydrogenation: Phosphoramidite libraries, ruthenacycles, and artificial enzymes. *Chem. Eur. J.* 2006, *12*, 4722–4734.
56. Reetz, M. T.; Mehler G.; Meiswinkel, A. Mixtures of chiral monodentate phosphites, phosphonites and phosphines as ligands in Rh-catalyzed hydrogenation of *N*-acyl enamines: Extension of the combinatorial approach. *Tetrahedron: Asymmetry* 2004, *15*, 2165–2167.
57. Jäkel C.; Paciello R. High-throughput and parallel screening methods in asymmetric hydrogenation. *Chem. Rev.* 2006, *106*, 2912–2942.
58. Reez, M. T.; Meiswinkel, A.; Mehler, G.; Angermund, K.; Graf, M.; Thiel, W.; Mynott R.; Blackmond, D. G. Why are BINOL-based monophosphites such efficient ligands in Rh-catalyzed asymmetric olefin hydrogenation? *J. Am. Chem. Soc.* 2005, *127*, 10305–10313.
59. Fu, Y.; Guo, X.-X.; Zhu, S.-F.; Hu, A. G.; Xie J.-H.; Zhu, Q.-L. Rhodium-catalyzed asymmetric hydrogenation of functionalized olefins using monodentate spiro phosphoramidite ligands. *J. Org. Chem.*, 2004, *69*, 4648–4655.
60. Gridnev, I. D.; Fan C.; Pringle, P. G. New insights into the mechanism of asymmetric hydrogenation catalysed by monophosphonite–rhodium complexes. *Chem. Commun.* 2007, *43*, 1319–1321.

61. Gridnev, I. D.; Alberico, E.; Gladiali, S. Captured at last: A catalyst–substrate adduct and a Rh-dihydride solvate in the asymmetric hydrogenation by a Rh-monophosphine catalyst. *Chem. Commun.* 2012, *48*, 2186–2188.

62. Jankowski, P.; McMullin, C. L.; Gridnev, I. D.; Orpen, A. G.; Pringle, P. G. Is restricted M–P rotation a common feature of enantioselective monophos catalysts? An example of restricted Rh–P rotation in a secondary phosphine complex. *Tetrahedron: Asymmetry* 2010, *21*, 1206–1209.

63. Noyori, R.; Ohkuma, T. Asymmetric catalysis by architectural and functional molecular engineering: Practical chemo- and stereoselective hydrogenation of ketones. *Angew. Chem. Int. Ed.* 2001, *40*, 40–73.

64. Wang, D.; Astruc, D. The golden age of transfer hydrogenation. *Chem. Rev.* 2015, *115*, 6621–6686.

65. Yoshimura, M.; Tanaka, S.; Kitamura, M. Recent topics in catalytic asymmetric hydrogenation of ketones. *Tetrahedron Lett.* 2014, *55*, 3635–3640.

66. Cotarca, L.; Verzini, M.; Volpicelli, R. Catalytic asymmetric transfer hydrogenation: An industrial perspective. *Chim. Oggi.* 2014, *32*, 36–41.

67. Stefane, B.; Pozgan, F. Advances in catalyst systems for the asymmetric hydrogenation and transfer hydrogenation of ketones. *Catal. Rev. Sci. Eng.* 2014, *56*, 82–174.

68. Václavík, J.; Šot, P.; Vilhanová, B.; Pecháček, J.; Kuzma, M.; Kačer, P. Practical aspects and mechanism of asymmetric hydrogenation with chiral half-sandwich complexes. *Molecules* 2013, *18*, 6804–6828.

69. Blaser, H.-U.; Pugin, B.; Spindler, F. Asymmetric hydrogenation. *Top. Organomet. Chem.* 2012, *42*, 65–102.

70. Magano, J.; Dunetz, J. R. Large-scale carbonyl reductions in the pharmaceutical industry. *Org. Process Res. Dev.* 2012, *16*, 1156–1184.

71. Ager, D. J.; de Vries, A. H. M.; de Vries, J. G. Asymmetric homogeneous hydrogenations at scale. *Chem. Soc. Rev.* 2012, *41*, 3340–3380.

72. Stefane, B.; Pozgan, F. Asymmetric hydrogenation and transfer hydrogenation of ketones. In *Hydrogenation*; Karamé, I., Ed.; InTech, 2012, pp. 31–68.

73. Darwish, M.; Wills, M. Asymmetric catalysis using iron complexes— 'Ruthenium Lite'? *Catal. Sci. Technol.* 2012, *2*, 243–255.

74. Ikariya, T. Chemistry of concerto molecular catalysis based on the metal/NH bifunctionality. *Bull. Chem. Soc. Jpn.* 2011, *84*, 1–16.

75. Václavík, J.; Kačer, P.; Kuzma, M.; Červený, L. Opportunities offered by chiral η6-arene/N-arylsulfonyl-diamine-RuII catalysts in the asymmetric transfer hydrogenation of ketones and imines. *Molecules* 2011, *16*, 5460–5495.

76. *Topics in Organometallic Chemistry: Bifunctional Molecular Catalysis*, Vol. 37, Ikariya, T.; Shibasaki, M., Eds.; Springer, New York, 2011.

77. Busacca, C. A.; Fandrick, D. R.; Song, J. J.; Senanayake, C. H. The growing impact of catalysis in the pharmaceutical industry. *Adv. Synth. Catal.* 2011, *353*, 1825–1864.

78. Ohkuma, T. Asymmetric hydrogenation of ketones: Tactics to achieve high reactivity, enantioselectivity, and wide scope. *Proc. Jpn. Acad. Ser. B Phys. Biol. Sci.* 2010, *86*, 202–219.

79. Wu, X.; Wang, C.; Xiao, J. Asymmetric transfer hydrogenation in water with platinum group metal catalysts. *Platinum Metals Rev.* 2010, *54*, 3–19.

80. Wu, X.; Xiao, J. Aqueous-phase asymmetric transfer hydrogenation of ketones – a greener approach to chiral alcohols. *Chem. Commun.* 2007, 2449–2466.

81. Shimizu, H.; Nagasaki, I.; Matsumura, K.; Sayo, N.; Saito, T. Developments in asymmetric hydrogenation from an industrial perspective. *Acc. Chem. Res.* 2007, *40*, 1385–1393.
82. Ikariya, T.; Blacker, A. J. Asymmetric transfer hydrogenation of ketones with bifunctional transition metal-based molecular. *Acc. Chem. Res.* 2007, *40*, 1300–1308.
83. Ikariya, T.; Murata, K.; Noyori, R. Bifunctional transition metal-based molecular catalysts for asymmetric syntheses. *Org. Biomol. Chem.* 2006, *4*, 393–406.
84. Muñiz, K. Bifunctional metal–ligand catalysis: Hydrogenations and new reactions within the metal–(di)amine scaffold. *Angew. Chem. Int. Ed.* 2005, *44*, 6622–6627.
85. Blaser, H. U.; Malan, C.; Pugin, B.; Spindler, F.; Steiner, H.; Studer, M. Selective hydrogenation for fine chemicals: Recent trends and new developments. *Adv. Synth. Catal.* 2003, *345*, 103–151.
86. Noyori, R.; Hashiguchi, S. Asymmetric transfer hydrogenation catalyzed by chiral ruthenium complexes. *Acc. Chem. Res.* 1997, *30*, 97–102.
87. Imamoto, T. Asymmetric hdrogenation In *Hydrogenation.* Karamé, I., Ed.; InTech, 2012.
88. Zheng, C.; You, S.-L. Transfer hydrogenation with Hantzsch esters and related organic hydride donors. *Chem. Soc. Rev.* 2012, *41*, 2498–2518.
89. Gladiali, S.; Alberico, E. Asymmetric transfer hydrogenation: Chiral ligands and applications. *Chem. Soc. Rev.* 2006, *35*, 226–236.
90. Zassinovich, G.; Mestroni, G.; Gladiali, S. Asymmetric hydrogen transfer reactions promoted by homogeneous transition metal catalysts. *Chem. Rev.* 1992, *92*, 1051–1069.
91. Bakos, J.; Tóth, I.; Heil, B.; Markó, L. A facile method for the preparation of 2,4-bis(diphenylphosphino)pentane (BDPP) enantiomers and their application in asymmetric hydrogenation. *J. Organomet. Chem.* 1985, *279*, 23–29.
92. Ohkuma, T.; Ooka, H.; Hashiguchi, S.; Ikariya, T.; Noyori, R. Practical enantioselective hydrogenation of aromatic ketones. *J. Am. Chem. Soc.* 1995, *117*, 2675–2676.
93. Noyori, R.; Sandoval, C. A.; Muniz, K.; Ohkuma, T. Metal-ligand bifunctional catalysis for asymmetric hydrogenation. *Philos. Trans. R. Soc. A-Math. Phys. Eng. Sci.* 2005, *363*, 901–912.
94. Noyori, R.; Kitamura, M.; Ohkuma, T. Toward efficient asymmetric hydrogenation: Architectural and functional engineering of chiral molecular catalysts. *Proc. Natl. Acad. Sci. U.S.A.* 2004, *101*, 5356–5362.
95. Noyori, R. Asymmetric catalysis: Science and opportunities (Nobel Lecture). *Angew. Chem., Int. Ed.* 2002, *41*, 2008–2022.
96. Noyori, R.; Yamakawa, M.; Hashiguchi, S. Metal–ligand bifunctional catalysis: A nonclassical mechanism for asymmetric hydrogen transfer between alcohols and carbonyl compounds. *J. Org. Chem.* 2001, *66*, 7931–7944.
97. Noyori, R.; Koizumi, M.; Ishii, D.; Ohkuma, T. Asymmetric hydrogenation via architectural and functional molecular engineering. *Pure Appl. Chem.* 2001, *73*, 227–232.
98. Emura, M.; Toyoda, T.; Kiyofuji, N.; Noyori, R.; Ikariya, T.; Okuma, T. Preparation of cis-4-tert-butylcyclohexanol from 4-tert-butylcyclohexanone JP09241194A.

99. England, A. F.; Ikariya, T.; Noyori, R. Trans-[RuCl2(phosphine)2(1, 2-diamine)] and chiral trans-[RuCl2(diphosphine)(1,2-diamine)]: Shelf-stable precatalysts for the rapid, productive, and stereoselective hydrogenation of ketones. *Angew. Chem. Int. Ed. Engl.* 1998, *37*, 1703–1707.
100. Hashiguchi, S.; Fujii, A.; Takehara, J.; Ikariya, T.; Noyori, R. Asymmetric transfer hydrogenation of aromatic ketones catalyzed by chiral Ruthenium(Ii) complexes. *J. Am. Chem. Soc.* 1995, *117*, 7562–7563.
101. Fujii, A.; Hashiguchi, S.; Uematsu, N.; Ikariya, T.; Noyori, R. Ruthenium(II)-catalyzed asymmetric transfer hydrogenation of ketones using a formic acid-triethylamine mixture. *J. Am. Chem. Soc.* 1996, *118*, 2521–2522.
102. Uematsu, N.; Fujii, A.; Hashiguchi, S.; Ikariya, T.; Noyori, R. Asymmetric transfer hydrogenation of imines. *J. Am. Chem. Soc.* 1996, *118*, 4916–4917.
103. Takehara, J.; Hashiguchi, S.; Fujii, A.; Inoue, S.-I.; Ikariya, T.; Noyori, R. Amino alcohol effects on the ruthenium(II)-catalysed asymmetric transfer hydrogenation of ketones in propan-2-ol. *Chem. Commun.* 1996, 233–234.
104. O'Neal, E. J.; Lee, C. H.; Brathwaite, J.; Jensen, K. F. Continuous nanofiltration and recycle of an asymmetric ketone hydrogenation catalyst. *ACS Cat.* 2015, *5*, 2615–2622.
105. Rodríguez, S.; Qu, B.; Fandrick, K. R.; Buono, F.; Haddad, N.; Xu, Y.; Herbage, M. A.; Zeng, X.; Ma, S.; Grinberg, N.; Lee, H.; Han, Z. S.; Yee, N. K.; Senanayake, C. H. Amine-tunable ruthenium catalysts for asymmetric reduction of ketones. *Adv. Synth. Catal.* 2014, *356*, 301–307.
106. Matsumura, K.; Arai, N.; Hori, K.; Saito, T.; Sayo, N.; Ohkuma, T. Chiral ruthenabicyclic complexes: Precatalysts for rapid, enantioselective, and wide-scope hydrogenation of ketones. *J. Am. Chem. Soc.* 2011, *133*, 10696–10699.
107. Li, Y.; Ding, K.; Sandoval, C. A. Hybrid NH2-benzimidazole ligands for efficient Ru-catalyzed asymmetric hydrogenation of aryl ketones. *Org. Lett.* 2009, *11*, 907–910.
108. Ooka, H.; Arai, N.; Azuma, K.; Kurono, N.; Ohkuma, T. Asymmetric hydrogenation of aromatic ketones catalyzed by the TolBINAP/ DMAPEN–Ruthenium(II) complex: A significant effect of N-substituents of chiral 1,2-diamine ligands on enantioselectivity. *J. Org. Chem.* 2008, *73*, 9084–9093.
109. Ohkuma, T.; Sandoval, C. A.; Srinivasan, R.; Lin, Q.; Wei, Y.; Muñiz, K.; Noyori, R. Asymmetric hydrogenation of tert-alkyl ketones. *J. Am. Chem. Soc.* 2005, *127*, 8288–8289.
110. Zimbron, J. M.; Dauphinais, M.; Charette, A. B. Noyori-Ikariya catalyst supported on tetra-arylphosphonium salt for asymmetric transfer hydrogenation in water. *Green Chem.* 2015, *17*, 3255–3259.
111. Soni, R.; Hall, T. H.; Morris, D. J.; Clarkson, G. J.; Owen, M. R.; Wills, M. N-functionalised TsDPEN catalysts for asymmetric transfer hydrogenation; synthesis and applications. *Tetrahedron Lett.* 2015, *56*, 6397–6401.
112. Soni, R.; Hall, T. H.; Mitchell, B. P.; Owen, M. R.; Wills, M. Asymmetric reduction of electron-rich ketones with tethered Ru(II)/TsDPEN catalysts using formic acid/triethylamine or aqueous sodium formate. *J. Org. Chem.* 2015, *80*, 6784–6793.
113. Jolley, K. E.; Clarkson, G. J.; Wills, M. Tethered Ru(II) catalysts containing a Ru-I bond. *J. Organomet. Chem.* 2015, *776*, 157–162.

114. Wills, M.; Jolley, K. E.; Soni, R. Catalyst and process for synthesising the same. WO2014068331A1, 2014.
115. Hodgkinson, R.; Jurcik, V.; Zanotti-Gerosa, A.; Nedden, H. G.; Blackaby, A.; Clarkson, G. J.; Wills, M. Synthesis and catalytic applications of an extended range of tethered ruthenium(II)/η6-arene/diamine complexes. *Organometallics* 2014, *33*, 5517–5524.
116. Fang, Z.; Wills, M. Asymmetric reduction of diynones and the total synthesis of (S)-panaxjapyne A. *Org. Lett.* 2014, *16*, 374–377.
117. Fang, Z.; Clarkson, G. J.; Wills, M. Asymmetric reduction of 2,2-dimethyl-6-(2-oxoalkyl/oxoaryl)-1,3-dioxin-4-ones and application to the synthesis of (+)-yashabushitriol. *Tetrahedron Lett.* 2013, *54*, 6834–6837.
118. Dub, P. A.; Wang, H.; Matsunami, A.; Gridnev, I. D.; Kuwata, S.; Ikariya, T. C-F bond breaking through aromatic nucleophilic substitution with a hydroxo ligand mediated via water bifunctional activation. *Bull. Chem. Soc. Jpn.* 2013, *86*, 557–568.
119. Parekh, V.; Ramsden, J. A.; Wills, M. Ether-tethered Ru(ii)/TsDPEN complexes; synthesis and applications to asymmetric transfer hydrogenation. *Catal. Sci. Technol.* 2012, *2*, 406–414.
120. Touge, T.; Hakamata, T.; Nara, H.; Kobayashi, T.; Sayo, N.; Saito, T.; Kayaki, Y.; Ikariya, T. Oxo-tethered ruthenium(II) complex as a bifunctional catalyst for asymmetric transfer hydrogenation and H2 hydrogenation. *J. Am. Chem. Soc.* 2011, *133*, 14960–14963.
121. Tang, W.; Johnston, S.; Li, C.; Iggo, J. A.; Bacsa, J.; Xiao, J. Cooperative catalysis: Combining an achiral metal catalyst with a chiral brønsted acid enables highly enantioselective hydrogenation of imines. *Chem.* 2013, *19*, 14187–14193.
122. Ward, T. R. Artificial metalloenzymes based on the biotin–avidin technology: Enantioselective catalysis and beyond. *Acc. Chem. Res.* 2010, *44*, 47–57.
123. Sun, X.; Gavriilidis, A. Scalable reactor design for pharmaceuticals and fine chemicals production. 3. A novel gas–liquid reactor for catalytic asymmetric transfer hydrogenation with simultaneous acetone stripping. *Org. Process Res. Dev.* 2008, *12*, 1218–1222.
124. Matharu, D. S.; Morris, D. J.; Clarkson, G. J.; Wills, M. An outstanding catalyst for asymmetric transfer hydrogenation in aqueous solution and formic acid/triethylamine. *Chem. Commun.* 2006, 3232–3234.
125. Blackmond, D. G.; Ropic, M.; Stefinovic, M. Kinetic studies of the asymmetric transfer hydrogenation of imines with formic acid catalyzed by Rh–diamine catalysts. *Org. Process Res. Dev.* 2006, *10*, 457–463.
126. Sun, X.; Manos, G.; Blacker, J.; Martin, J.; Gavriilidis, A. Asymmetric transfer hydrogenation of acetophenone with 1R,2S-aminoindanol/pentamethylcyclopentadienylrhodium Catalyst. *Org. Process Res. Dev.* 2004, *8*, 909–914.
127. Mao, J.; Baker, D. C. A chiral rhodium complex for rapid asymmetric transfer hydrogenation of imines with high enantioselectivity. *Org. Lett.* 1999, *1*, 841–843.
128. Murata, K.; Ikariya, T.; Noyori, R. New chiral rhodium and iridium complexes with chiral diamine ligands for asymmetric transfer hydrogenation of aromatic ketones. *J. Org. Chem.* 1999, *64*, 2186–2187.
129. Mashima, K.; Abe, T.; Tani, K. Asymmetric transfer hydrogenation of ketonic substrates catalyzed by (eta(5)-C5Me5)MCl complexes (M = Rh and Ir) of (1S,2S)-N-(p-toluenesulfonyl)-1,2-diphenylethylenediamine. *Chem. Lett.* 1998, 1199–1200.

130. Ohkuma, T.; Utsumi, N.; Watanabe, M.; Tsutsumi, K.; Arai, N.; Murata, K. Asymmetric hydrogenation of α-hydroxy ketones catalyzed by MsDPEN–Cp*Ir(III) complex. *Org. Lett.* 2007, *9*, 2565–2567.

131. Xu, L.; Huang, Z.-H.; Sandoval, C. A.; Gu, L.-Q.; Huang, Z.-S. Asymmetric hydrogenation of 3,5-bistrifluoromethyl acetophenone in pilot scale with industrially viable Ru/diphosphine–benzimidazole complexes. *Org. Process Res. Dev.* 2014, *18*, 1137–1141.

132. Yan, P.-C.; Zhu, G.-L.; Xie, J.-H.; Zhang, X.-D.; Zhou, Q.-L.; Li, Y.-Q.; Shen, W.-H.; Che, D.-Q. Industrial scale-up of enantioselective hydrogenation for the asymmetric synthesis of rivastigmine. *Org. Process Res. Dev.* 2013, *17*, 307–312.

133. Verzijl, G. K. M.; de Vries, A. H. M.; de Vries, J. G.; Kapitan, P.; Dax, T.; Helms, M.; Nazir, Z.; Skranc, W.; Imboden, C.; Stichler, J.; Ward, R. A.; Abele, S.; Lefort, L. Catalytic asymmetric reduction of a 3,4-dihydroisoquinoline for the large-scale production of almorexant: Hydrogenation or transfer hydrogenation? *Org. Process Res. Dev.* 2013, *17*, 1531–1539.

134. Cheng, J.-J.; Yang, Y.-S. Enantioselective total synthesis of (–)-(S)-stepholidine. *J. Org. Chem.* 2009, *74*, 9225–9228.

135. Zhang, J.; Blazecka, P. G.; Bruendl, M. M.; Huang, Y. Ru-TsDPEN with formic acid/Hünig's base for asymmetric transfer hydrogenation, a practical synthesis of optically enriched N-propyl pantolactam. *J. Org. Chem.* 2009, *74*, 1411–1414.

136. Sharma, P. K.; Kolchinski, A.; Shea, H. A.; Nair, J. J.; Gou, Y.; Romanczyk, L. J.; Schmitz, H. H. Scale-up syntheses of two naturally occurring procyanidins: (–)-epicatechin-(4β,8)-(+)-catechin and (–)-Epicatechin-3-O-galloyl-(4β,8)-(–)-epicatechin-3-O-gallate. *Org. Process Res. Dev.* 2007, *11*, 422–430.

137. Chen, C.-Y.; Frey, L. F.; Shultz, S.; Wallace, D. J.; Marcantonio, K.; Payack, J. F.; Vazquez, E.; Springfield, S. A.; Zhou, G.; Liu, P.; Kieczykowski, G. R.; Chen, A. M.; Phenix, B. D.; Singh, U.; Strine, J.; Izzo, B.; Krska, S. W. Catalytic, enantioselective synthesis of taranabant, a novel, acyclic cannabinoid-1 receptor inverse agonist for the treatment of obesity. *Org. Process Res. Dev.* 2007, *11*, 616–623.

138. Li, X.; Wu, X.; Chen, W.; Hancock, F. E.; King, F.; Xiao, J. Asymmetric transfer hydrogenation in water with a supported Noyori–Ikariya catalyst. *Org. Lett.* 2004, *6*, 3321–3324.

139. Hansen, K. B.; Chilenski, J. R.; Desmond, R.; Devine, P. N.; Grabowski, E. J. J.; Heid, R.; Kubryk, M.; Mathre, D. J.; Varsolona, R. Scalable, efficient process for the synthesis of (R)-3,5-bistrifluoromethylphenyl ethanol via catalytic asymmetric transfer hydrogenation and isolation as a DABCO inclusion complex. *Tetrahedron: Asymmetry* 2003, *14*, 3581–3587.

140. Brands, K. M. J.; Payack, J. F.; Rosen, J. D.; Nelson, T. D.; Candelario, A.; Huffman, M. A.; Zhao, M. M.; Li, J.; Craig, B.; Song, Z. J.; Tschaen, D. M.; Hansen, K.; Devine, P. N.; Pye, P. J.; Rossen, K.; Dormer, P. G.; Reamer, R. A.; Welch, C. J.; Mathre, D. J.; Tsou, N. N.; McNamara, J. M.; Reider, P. J. Efficient synthesis of NK1 receptor antagonist aprepitant using a crystallization-induced diastereoselective transformation. *J. Am. Chem. Soc.* 2003, *125*, 2129–2135.

141. Miyagi, M.; Takehara, J.; Collet, S.; Okano, K. Practical synthesis of (S)-1-(3-trifluoromethylphenyl)ethanol via ruthenium(II)-catalyzed asymmetric transfer hydrogenation. *Org. Process Res. Dev.* 2000, *4*, 346–348.

142. Yang, H.; Huo, N.; Yang, P.; Pei, H.; Lv, H.; Zhang, X. Rhodium catalyzed asymmetric hydrogenation of 2-pyridine ketones. *Org. Lett.* 2015, *17*, 4144–4147.

143. Shang, G.; Liu, D.; Allen, S. E.; Yang, Q.; Zhang, X. Asymmetric hydrogenation of α-primary and secondary amino ketones: Efficient asymmetric syntheses of (–)-Arbutamine and (–)-Denopamine. *Chem.* 2007, *13*, 7780–7784.
144. Jiang, Q.; Jiang, Y.; Xiao, D.; Cao, P.; Zhang, X. Highly enantioselective hydrogenation of simple ketones catalyzed by a Rh–PennPhos complex. *Angew. Chem. Int. Ed.* 1998, *37*, 1100–1103.
145. Zhang, W.; Chi, Y.; Zhang, X. Developing chiral ligands for asymmetric hydrogenation. *Acc. Chem. Res.* 2007, *40*, 1278–1290.
146. Chi, Y.; Tang, W.; Zhang, X. Rhodium-catalyzed asymmetric hydrogenation. In *Modern Rhodium-Catalyzed Organic Reactions*; Wiley-VCH, 2005, pp. 1–31.
147. Zhao, B.; Han, Z.; Ding, K. The N-H functional group in organometallic catalysis. *Angew. Chem. Int. Ed.* 2013, *52*, 4744–4788.
148. Nara, H.; Yokozawa, T. Ruthenium complex and method for preparing optically active alcohol compound WO2011135753A1, 2011.
149. Xie, J.-H.; Liu, X.-Y.; Xie, J.-B.; Wang, L.-X.; Zhou, Q.-L. An additional coordination group leads to extremely efficient chiral iridium catalysts for asymmetric hydrogenation of ketones. *Angew. Chem. Int. Ed.* 2011, *50*, 7329–7332.
150. Wu, W.; Liu, S.; Duan, M.; Tan, X.; Chen, C.; Xie, Y.; Lan, Y.; Dong, X.-Q.; Zhang, X. Iridium catalysts with f-amphox ligands: Asymmetric hydrogenation of simple ketones. *Org. Lett.* 2016, *18*, 2938–2941.
151. Kuwata, S.; Ikariya, T. Metal-ligand bifunctional reactivity and catalysis of protic N-heterocyclic carbene and pyrazole complexes featuring [small beta]-NH units. *Chem. Commun.* 2014, *50*, 14290–14300.
152. Wang, C.; Xiao, J. Asymmetric reductive amination. *Top. Curr. Chem.* 2014, *343*, 261–282.
153. Václavík, J.; Šot, P.; Pecháček, J.; Vilhanová, B.; Matuška, O.; Kuzma, M.; Kačer, P. Experimental and theoretical perspectives of the Noyori-Ikariya asymmetric transfer hydrogenation of imines. *Molecules* 2014, *19*, 6987–7007.
154. Xu, L.; Wu, X.; Xiao, J. Stereoselective reduction of imino groups. In *Stereoselective Synthesis 2: Stereoselective Reactions of Carbonyl and Imino Groups*, Vol. 2, Molander, G. A., Ed.; Georg Thieme Verlag, 2011, pp. 251–309.
155. Wang, C.; Villa-Marcos, B.; Xiao, J. Hydrogenation of imino bonds with half-sandwich metal catalysts. *Chem. Commun.* 2011, *47*, 9773–9785.
156. Fleury-Bregeot, N.; de la Fuente, V.; Castillon, S.; Claver, C. Highlights of transition metal-catalyzed asymmetric hydrogenation of imines. *ChemCatChem* 2010, *2*, 1346–1371.
157. Church, T. L.; Andersson, P. G. *Chiral Amines from Transition-Metal-Mediated Hydrogenation and Transfer Hydrogenation.* Wiley-VCH, 2010, pp. 179–223.
158. Nugent, T. C.; El-Shazly, M. Chiral amine synthesis—Recent developments and trends for enamide reduction, reductive amination, and imine reduction. *Adv. Synth. Catal.* 2010, *352*, 753–819.
159. Blaser, H.-U.; Spindler, F. Catalytic asymmetric hydrogenation of C=N functions. *Org. React. (Hoboken, NJ, U. S.)* 2009, *74*, 1–102.
160. Dub, P. A.; Ikariya, T. Catalytic reductive transformations of carboxylic and carbonic acid derivatives using molecular hydrogen. *ACS Cat.* 2012, *2*, 1718–1741.
161. Ito, M.; Ikariya, T. Catalytic hydrogenation of polar organic functionalities based on Ru-mediated heterolytic dihydrogen cleavage. *Chem. Commun.* 2007, 5134–5142.

162. Pritchard, J.; Filonenko, G. A.; van Putten, R.; Hensen, E. J. M.; Pidko, E. A. Heterogeneous and homogeneous catalysis for the hydrogenation of carboxylic acid derivatives: History, advances and future directions. *Chem. Soc. Rev.* 2015, *44*, 3808–3833.

163. Mérel, D. S.; Do, M. L. T.; Gaillard, S.; Dupau, P.; Renaud, J.-L. Iron-catalyzed reduction of carboxylic and carbonic acid derivatives. *Coord. Chem. Rev.* 2015, *288*, 50–68.

164. Ikariya, T.; Kayaki, Y. Hydrogenation of carboxylic acid derivatives with bifunctional ruthenium catalysts. *Pure Appl. Chem.* 2014, *86*, 933–943.

165. Werkmeister, S.; Junge, K.; Beller, M. Catalytic hydrogenation of carboxylic acid esters, amides, and nitriles with homogeneous catalysts. *Org. Process Res. Dev.* 2014, *18*, 289–302.

166. Clarke, M. L. Recent developments in the homogeneous hydrogenation of carboxylic acid esters. *Catal. Sci. Technol.* 2012, *2*, 2418–2423.

167. Werkmeister, S.; Neumann, J.; Junge, K.; Beller, M. Pincer-type complexes for catalytic (de)hydrogenation and transfer (de)hydrogenation reactions: Recent progress. *Chem.* 2015, *21*, 12226–12250.

168. Friedrich, A.; Schneider, S. Acceptorless dehydrogenation of alcohols: Perspectives for synthesis and H2 storage. *ChemCatChem* 2009, *1*, 72–73.

169. Ikariya, T.; Gridnev, I. D. Bifunctional transition metal-based molecular catalysts for asymmetric C–C and C–N bond formation. *Top. Catal.* 2010, *53*, 894–901.

170. Ikariya, T.; Gridnev, I. D. Bifunctional transition metal-based molecular catalysts for asymmetric C–C and C–N bond formation. *Chem. Rec.* 2009, *9*, 106–123.

171. Heisenberg, W. Über den anschaulichen Inhalt der quantentheoretischen Kinematik und Mechanik. *Z. Physik* 1927, *43*, 172–198.

172. Dewar, M. J. S. Multibond reactions cannot normally be synchronous. *J. Am. Chem. Soc.* 1984, *106*, 209–219.

173. Tantillo, D. J. Recent excursions to the borderlands between the realms of concerted and stepwise: Carbocation cascades in natural products biosynthesis. *J. Phys. Org. Chem.* 2008, *21*, 561–570.

174. Jones II, G. Putative diradicals. *J. Chem. Educ.* 1974, *51*, 175.

175. Marvet, U.; Dantus, M. Femtosecond observation of a concerted chemical reaction. *Chem. Phys. Lett.* 1996, *256*, 57–62.

176. Strauss, C. E. M.; Houston, P. L. Correlations without coincidence measurements: Deciding between stepwise and concerted dissociation mechanisms for ABC.-+. A + B + C. *J. Phys. Chem.* 1990, *94*, 8751–8762.

177. Borden, W. T.; Loncharich, R. J.; Houk, K. N. Synchronicity in multibond reactions. *Annu. Rev. Phys. Chem.* 1988, *39*, 213–236.

178. Woodward, R. B.; Hoffmann, R. The conservation of orbital symmetry. *Angew. Chem. Int. Ed.* 1969, *8*, 781–853.

179. Polanyi, J. C.; Zewail, A. H. Direct observation of the transition state. *Acc. Chem. Res.* 1995, *28*, 119–132.

180. Kim, S. K.; Pedersen, S.; Zewail, A. H. Direct femtosecond observation of the transient intermediate in the alpha-cleavage reaction of (Ch3)(2)Co To 2ch(3)+Co—Resolving the issue of concertedness. *J. Chem. Phys.* 1995, *103*, 477–480.

181. Pedersen, S.; Herek, J. L.; Zewail, A. H. The validity of the "diradical" hypothesis: Direct femtosecond studies of the transition-state structures. *Science* 1994, *266*, 1359–1364.

182. Thomas, J. M.; Zamaraev, K. I. *Perspectives in Catalysis,* Blackwell Scientific Publications, Oxford, 1992.

183. Steinfeld, J. I.; Francisco, J. S.; Hase, W. L. *Chemical Kinetics and Dynamics,* 2nd ed., Prentice Hall, Upper Saddle River, NJ, 1998.

184. Wigner, E. The transition-state method. *Trans. Faraday Soc.* 1938, *34,* 29–41.

185. Eyring, H. Statistical mechanical treatment of the activated complex in chemical reactions. *J. Chem. Phys.* 1935, *3,* 107–115.

186. Marcus, R. A. Lifetimes of active molecules. I. *J. Chem. Phys.* 1952, *20,* 352–354.

187. Marcus, R. A. Lifetimes of active molecules. II. *J. Chem. Phys.* 1952, *20,* 355–359.

188. Kassel, L. S. Dynamics of unimolecular reactions. *Chem. Rev.* 1932, *10,* 11–25.

189. Rice, O. K.; Ramsperger, H. C. Theories of unimolecular gas reactions at low pressures. *J. Am. Chem. Soc.* 1927, *49,* 1617–1629.

190. Carpenter, B. K. Energy disposition in reactive intermediates. *Chem. Rev.* 2013, *113,* 7265–7286.

191. Rehbein, J.; Carpenter, B. K. Do we fully understand what controls chemical selectivity? *Phys. Chem. Chem. Phys.* 2011, *13,* 20906–20922.

192. Carpenter, B. K. Intramolecular dynamics for the organic chemist. *Acc. Chem. Res.* 1992, *25,* 520–528.

193. Wu, X.; Liu, J.; Di, T. D.; Iggo, J. A.; Catlow, C. R. A.; Bacsa, J.; Xiao, J. A multi-lateral mechanistic study into asymmetric transfer hydrogenation in water. *Chem.* 2008, *14,* 7699–7715.

194. Bell, R. P. Liversidge lecture. Recent advances in the study of kinetic hydrogen isotope effects. *Chem. Soc. Rev.* 1974, *3,* 513–544.

195. Quapp, W. How does a reaction path branching take place? A classification of bifurcation events. *J. Mol. Struct.* 2004, *695,* 95–101.

196. Ess, D. H.; Wheeler, S. E.; Iafe, R. G.; Xu, L.; Celebi-Oelcuem, N.; Houk, K. N. Bifurcations on potential energy surfaces of organic reactions. *Angew. Chem., Int. Ed.* 2008, *47,* 7592–7601.

197. Minyaev, R. M.; Getmanskii, I. V.; Quapp, W. A second-order saddle point in the reaction coordinate for the isomerization of the NH5 complex: Ab initio calculations. *Russ. J. Phys. Chem* 2004, *78,* 1494–1498.

198. Hirsch, M.; Quapp, W.; Heidrich, D. The set of valley-ridge inflection points on the potential energy surface of water. *Phys. Chem. Chem. Phys.* 1999, *1,* 5291–5299.

199. Quapp, W.; Heidrich, D. Analysis of the concept of minimum energy path on the potential-energy surface of chemically reacting systems. *Theor. Chim. Acta* 1984, *66,* 245–260.

200. Boudart, M. From the century of the rate equation to the century of the rate constants: A revolution in catalytic kinetics and assisted catalyst design. *Catal. Lett.* 2000, *65,* 1–3.

201. Hartmann, R.; Chen, P. Numerical modeling of differential kinetics in the asymmetric hydrogenation of acetophenone by Noyori's catalyst. *Adv. Synth. Catal.* 2003, *345,* 1353–1359.

202. Dumesic, J. A.; Rudd, D. F.; Aparicio, L. M.; Rekoske, J. E.; Trevino, A. A. *The Microkinetics of Heterogeneous Catalysis,* American Chemical Society, Washington, DC, 1993.
    Yaluris, G.; Rekoske, J. E.; Aparicio, L. M.; Madon, R. J.; Dumesic, J. A.; *J. Catal.* 1995, *153,* 65–75.
    Cortright, R. D.; Dumesic; J. A.; Madon, R. J.; *Topics in Catal.* 1997, *4,* 15–26.

203. Kozuch, S. Is there something new under the sun? myths and facts in the analysis of catalytic cycles. In *Understanding Organometallic Reaction Mechanisms and Catalysis*, Ananikov, V. P., Ed.; Wiley-VCH Verlag GmbH & Co. KGaA, Weinheim, Germany, 2015, pp. 217–247.

204. Kozuch, S. Steady state kinetics of any catalytic network: Graph theory, the energy span model, the analogy between catalysis and electrical circuits, and the meaning of "mechanism". *ACS Catal.* 2015, *5*, 5242–5255.

205. Kozuch, S. A refinement of everyday thinking: The energetic span model for kinetic assessment of catalytic cycles. *Wiley Interdiscip. Rev. Comput. Mol. Sci.* 2012, *2*, 795–815.

206. Kozuch, S.; Martin, J. M. L. The rate-determining step is dead. Long live the rate-determining state! *ChemPhysChem* 2011, *12*, 1413–1418.

207. Uhe, A.; Kozuch, S.; Shaik, S. Automatic analysis of computed catalytic cycles. *J. Comput. Chem.* 2011, *32*, 978–985.

208. Kozuch, S.; Shaik, S. How to conceptualize catalytic cycles? The energetic span model. *Acc. Chem. Res.* 2010, *44*, 101–110.

209. Kozuch, S. Reply to comment on '"Turning Over' definitions in catalytic cycles". *ACS Cat.* 2013, *3*, 380–380.

210. Lente, G. Comment on '"Turning Over' definitions in catalytic cycles". *ACS Cat.* 2013, *3*, 381–382.

211. Kozuch, S.; Martin, J. M. L. "Turning Over" definitions in catalytic cycles. *ACS Cat.* 2012, *2*, 2787–2794.

212. Herrmann, W. A.; Cornils, B. Organometallic homogeneous catalysis—Quo vadis? *Angew. Chem., Int. Ed.* 1997, *36*, 1048–1067.

213. Zuo, W.; Tauer, S.; Prokopchuk, D. E.; Morris, R. H. Iron catalysts containing amine(imine)diphosphine P-NH-N-P ligands catalyze both the asymmetric hydrogenation and asymmetric transfer hydrogenation of ketones. *Organometallics* 2014, *33*, 5791–5801.

214. Abdur-Rashid, K.; Clapham, S. E.; Hadzovic, A.; Harvey, J. N.; Lough, A. J.; Morris, R. H. Mechanism of the hydrogenation of ketones catalyzed by trans-dihydrido(diamine)ruthenium(II) complexes. *J. Am. Chem. Soc.* 2002, *124*, 15104–15118.

215. Dudnik, A. S.; Weidner, V. L.; Motta, A.; Delferro, M.; Marks, T. J. Atom-efficient regioselective 1,2-dearomatization of functionalized pyridines by an earth-abundant organolanthanide catalyst. *Nat. Chem.* 2014, *6*, 1100–1107.

216. Kozuch, S.; Lee, S. E.; Shaik, S. Theoretical analysis of the catalytic cycle of a Nickel cross-coupling process: Application of the energetic span model. *Organometallics* 2009, *28*, 1303–1308.

217. Different approach of the simulation example of the simulation of kinetics for the computed catalytic cycle:
    Gridnev, I. D.; Vorobiev, A. K. *Bull. Chem. Soc. Jpn.* 2015, *88*, 333.
    Gridnev, I. D.; Vorobiev, A. K. *ACS Catal.* 2012, *2*, 2137.

218. Kohn, W.; Sham, L. J. Self-consistent equations including exchange and correlation effects. *Phys. Rev.* 1965, *140*, A1133–A1138.

219. Hohenberg, P.; Kohn, W. Inhomogeneous electron gas. *Phys. Rev.* 1964, *136*, B864–B871.

220. Reichardt, C.; Welton, T. *Solvents and Solvent Effects in Organic Chemistry*, 4th ed., Wiley-VCH, Weinheim, Germany, 2010.

221. Monard, G.; Rivail, J.-L. Solvent effects in quantum chemistry. In *Handbook of Computational Chemistry*, Leszczynski, J., Ed.; Springer, the Netherlands, 2014, pp. 561–571.
222. Tapia, O.; Bertrán, J. *Solvent Effects and Chemical Reactivity*, Springer, the Netherlands, 1996.
223. Rivail, J. L.; Rinaldi, D.; Ruiz-Lopez, M. F. *In Liquid State Quantum Chemistry in Computational Chemistry: Review of Current Trends*, World Scientific, Singapore, 1995, p. 65.
224. Clay, D. R.; McIntosh, M. C. Anomalies in the asymmetric transfer hydrogenation of several polycyclic meso compounds. *Tetrahedron Lett.* 2012, *53*, 1691–1694.
225. Arseniyadis, S.; Valleix, A.; Wagner, A.; Mioskowski, C. Kinetic resolution of amines: A highly enantioselective and chemoselective acetylating agent with a unique solvent-induced reversal of stereoselectivity. *Angew. Chem. Int. Ed.* 2004, *43*, 3314–3317.
226. Karelson, M. M.; Katritzky, A. R.; Zerner, M. C. Reaction field effects on the electron distribution and chemical reactivity of molecules. *Int. J. Quantum Chem.* 1986, *30*, 521–527.
227. Lipparini, F.; Scalmani, G.; Lagardere, L.; Stamm, B.; Cances, E.; Maday, Y.; Piquemal, J.-P.; Frisch, M. J.; Mennucci, B. Quantum, classical, and hybrid QM/MM calculations in solution: General implementation of the ddCOSMO linear scaling strategy. *J. Chem. Phys.* 2014, *141*, 184108.
228. Cramer, C. J.; Truhlar, D. G. A universal approach to solvation modeling. *Acc. Chem. Res.* 2008, *41*, 760–768.
229. Mennucci, B.; Cammi, R. *Continuum Solvation Models in Chemical Physics: From Theory to Applications*; John Wiley & Sons, Chichester, England, 2007.
230. Tomasi, J.; Mennucci, B.; Cammi, R. Quantum mechanical continuum solvation models. *Chem. Rev.* 2005, *105*, 2999–3094.
231. Orozco, M.; Luque, F. J. Theoretical methods for the description of the solvent effect in biomolecular systems. *Chem. Rev.* 2000, *100*, 4187–4226.
232. Cramer, C. J.; Truhlar, D. G. Implicit solvation models: Equilibria, structure, spectra, and dynamics. *Chem. Rev.* 1999, *99*, 2161–2200.
233. Tomasi, J.; Persico, M. Molecular-interactions in solution—An overview of methods based on continuous distributions of the solvent. *Chem. Rev.* 1994, *94*, 2027–2094.
234. Feig, M. *Modeling Solvent Environments: Applications to Simulations of Biomolecules*, Wiley, Chichester, 2010.
235. Marenich, A. V.; Ding, W. D.; Cramer, C. J.; Truhlar, D. G. Resolution of a challenge for solvation modeling: Calculation of dicarboxylic acid dissociation constants using mixed discrete-continuum solvation models. *J. Phys. Chem. Lett.* 2012, *3*, 1437–1442.
236. Kovacs, G.; Rossin, A.; Gonsalvi, L.; Lledos, A.; Peruzzini, M. Comparative DFT analysis of ligand and solvent effects on the mechanism of H(2) activation in water mediated by half-sandwich complexes Cp 'Ru(PTA)(2)Cl (Cp '= C(5)H(5), C(5)Me(5); PTA=1,3,5-triaza-7-phosphaadamantane). *Organometallics* 2010, *29*, 5121–5131.
237. Lee, M. S.; Salsbury, F. R.; Olson, M. A. An efficient hybrid explicit/implicit solvent method for biomolecular simulations. *J. Comp. Chem.* 2004, *25*, 1967–1978.

238. Martinez, J. M.; Pappalardo, R. R.; Marcos, E. S. Study of the Ag+ hydration by means of a semicontinuum quantum-chemical solvation model. *J. Phys. Chem. A* 1997, *101*, 4444–4448.

239. Luque, F. J.; Orozco, M. Semiclassical-continuum approach to the electrostatic free energy of solvation. *J. Phys. Chem. B* 1997, *101*, 5573–5582.

240. Orr-Ewing, A. J. Reaction mechanisms: Stripping down SN2. *Nat. Chem.* 2012, *4*, 522–523.

241. Otto, R.; Brox, J.; Trippel, S.; Stei, M.; Best, T.; Wester, R. Single solvent molecules can affect the dynamics of substitution reactions. *Nat. Chem.* 2012, *4*, 534–538.

242. Riveros, J. M.; Jose, S. M.; Takashima, K. Gas-phase nucleophilic displacement-reactions. *Adv. Phys. Org. Chem.* 1985, *21*, 197–240.

243. Vidossich, P.; Lledós, A.; Ujaque, G.: *Realistic Simulation of Organometallic Reactivity in Solution by Means of First-Principles Molecular Dynamics (in Structure and Bonding)*, Springer, Berlin, Germany, 2015, p. 183.

244. Sunoj, R. B.; Anand, M. Microsolvated transition state models for improved insight into chemical properties and reaction mechanisms. *Phys. Chem. Chem. Phys.* 2012, *14*, 12715–12736.

245. Karaiskos, C. S.; Matiadis, D.; Markopoulos, J.; Igglessi-Markopoulou, O. Homogeneous chemoselective hydrogenation of heterocyclic compounds – The case of 1,4 addition on conjugated C-C and C-O double bonds of arylidene tetramic acids. In *Hydrogenation*; Karamé, I., Ed.; InTech, 2012.

246. Clapham, S. E.; Hadzovic, A.; Morris, R. H. Mechanisms of the H2-hydrogenation and transfer hydrogenation of polar bonds catalyzed by ruthenium hydride complexes. *Coord. Chem. Rev.* 2004, *248*, 2201–2237.

247. Salvini, A.; Frediani, P.; Gallerini, S. Homogeneous hydrogenation of ketones in the presence of H2Ru(CO)2(PPh3)2. *Appl. Organomet. Chem.* 2000, *14*, 570–580.

248. Joó, F.; Kovács, J.; Bényei, A. C.; Kathó, Á. The effects of pH on the molecular distribution of water soluble ruthenium(II) hydrides and its consequences on the selectivity of the catalytic hydrogenation of unsaturated aldehydes. *Catal. Today* 1998, *42*, 441–448.

249. Faza, O. N.; López, C. S.; Fernández, I. Noyori hydrogenation: Aromaticity, synchronicity, and activation strain analysis. *J. Org. Chem.* 2013, *78*, 5669–5676.

250. Although, it does not explain why there is some catalytic activity for fully alkylated catalysts.

251. Khusnutdinova, J. R.; Milstein, D. Metal–ligand cooperation. *Angew. Chem. Int. Ed.* 2015, *54*, 12236–12273.

252. Morris, R. H. Exploiting metal–ligand bifunctional reactions in the design of iron asymmetric hydrogenation catalysts. *Acc. Chem. Res.* 2015, *48*, 1494–1502.

253. Dub, P. A.; Henson, N. J.; Martin, R. L.; Gordon, J. C. Unravelling the mechanism of the asymmetric hydrogenation of acetophenone by [RuX2(diphosphine) (1,2-diamine)] catalysts. *J. Am. Chem. Soc.* 2014, *136*, 3505–3521.

254. Dub, P. A.; Ikariya, T. Quantum chemical calculations with the inclusion of nonspecific and specific solvation: Asymmetric transfer hydrogenation with bifunctional ruthenium catalysts. *J. Am. Chem. Soc.* 2013, *135*, 2604–2619.

255. In order to distinguish confusion from redox processes, see Refs. [256–259], the term "chemically noninnocent" or "chemically innocent ligand" is used to state the ability of a ligand to participate or not in the bond breaking/

making events, respectively. The oxidation state of the metal and the electronic structure of the ligand is well-defined and can be a priori unambiguously determined.

256. Caulton, K. G. Systematics and future projections concerning redox-noninnocent amide/imine ligands. *Eur. J. Inorg. Chem.* 2012, *2012*, 435–443.

257. Lyaskovskyy, V.; de Bruin, B. Redox non-innocent ligands: Versatile new tools to control catalytic reactions. *ACS Cat.* 2012, *2*, 270–279.

258. Kim, P. B.; Elena, K. B.; Nikolai, V. Z. Metal complexes with non-innocent ligands. *Russ. Chem. Rev.* 2005, *74*, 531.

259. Jørgensen, C. K. Differences between the four halide ligands, and discussion remarks on trigonal-bipyramidal complexes, on oxidation states, and on diagonal elements of one-electron energy. *Coord. Chem. Rev.* 1966, *1*, 164–178.

260. Sandoval, C. A.; Shi, Q.; Liu, S.; Noyori, R. NH/π Attraction: A role in asymmetric hydrogenation of aromatic ketones with binap/1,2-diamine-ruthenium(II) complexes. *Chem.* 2009, *4*, 1221–1224.

261. Sandoval, C. A.; Yamaguchi, Y.; Ohkuma, T.; Kato, K.; Noyori, R. Solution structures and behavior of trans-RuH(eta(1)-BH4) (binap)(1,2-diamine) complexes. *Magn. Reson. Chem.* 2006, *44*, 66–75.

262. Sandoval, C. A.; Ohkuma, T.; Muniz, K.; Noyori, R. Mechanism of asymmetric hydrogenation of ketones catalyzed by BINAP/1,2-diamine-ruthenium(II) complexes. *J. Am. Chem. Soc.* 2003, *125*, 13490–13503.

263. Casey, C. P.; Johnson, J. B. Kinetic isotope effect evidence for a concerted hydrogen transfer mechanism in transfer hydrogenations catalyzed by [p-(Me2CH)C6H4Me]Ru(NHCHPhCHPhNSO2C6H4-p-CH3). *J. Org. Chem.* 2003, *68*, 1998–2001.

264. Haack, K.-J.; Hashiguchi, S.; Fujii, A.; Ikariya, T.; Noyori, R. The catalyst precursor, catalyst and intermediate in the RuII-promoted asymmetric hydrogen transfer between alcohols and ketones. *Angew. Chem., Int. Ed. Engl.* 1997, *36*, 285–288.

265. Limé, E.; Lundholm, M. D.; Forbes, A.; Wiest, O.; Helquist, P.; Norrby, P.-O. Stereoselectivity in asymmetric catalysis: The case of ruthenium-catalyzed ketone hydrogenation. *J. Chem. Theory Comput.* 2014, *10*, 2427–2435.

266. Faza, O. N.; Fernandez, I.; Lopez, C. S. Computational insights on the mechanism of the catalytic hydrogenation with BINAP-diamine-Ru complexes: The role of base and origin of selectivity. *Chem. Commun.* 2013, *49*, 4277–4279.

267. Feng, R.; Xiao, A.; Zhang, X.; Tang, Y.; Lei, M. Origins of enantioselectivity in asymmetric ketone hydrogenation catalyzed by a RuH2(binap)(cydn) complex: Insights from a computational study. *Dalton Trans.* 2013, *42*, 2130–2145.

268. Chen, H.-Y. T.; Di, T. D.; Hogarth, G.; Catlow, C. R. A. The effects of ligand variation on enantioselective hydrogenation catalysed by RuH2(diphosphine) (diamine) complexes. *Dalton Trans.* 2012, *41*, 1867–1877.

269. Zhang, X.; Guo, X.; Chen, Y.; Tang, Y.; Lei, M.; Fang, W. Mechanism investigation of ketone hydrogenation catalyzed by ruthenium bifunctional catalysts: Insights from a DFT study. *Phys. Chem. Chem. Phys.* 2012, *14*, 6003–6012.

270. Chen, H.-Y. T.; Di, T. D.; Hogarth, G.; Catlow, C. R. A. Correlating enantioselectivity with activation energies in the asymmetric hydrogenation of acetophenone catalyzed by Noyori-type complexes. *Catal. Lett.* 2011, *141*, 1761–1766.

271. Chen, H.-Y. T.; Di, T. D.; Hogarth, G.; Catlow, C. R. A. trans-FeII(H)2(diphosphine)(diamine) complexes as alternative catalysts for the asymmetric hydrogenation of ketones? A DFT study. *Dalton Trans.* 2011, *40*, 402–412.
272. Zimmer-De, I. M.; Morris, R. H. Kinetic hydrogen/deuterium effects in the direct hydrogenation of ketones catalyzed by a well-defined ruthenium diphosphine diamine complex. *J. Am. Chem. Soc.* 2009, *131*, 11263–11269.
273. Chen, Y.; Tang, Y.; Lei, M. A comparative study on the hydrogenation of ketones catalyzed by diphosphine-diamine transition metal complexes using DFT method. *Dalton Trans.* 2009, 2359–2364.
274. Di Tommaso, D.; French, S. A.; Zanotti-Gerosa, A.; Hancock, F.; Palin, E. J.; Catlow, C. R. A. Computational study of the factors controlling enantioselectivity in ruthenium(II) hydrogenation catalysts. *Inorg. Chem.* 2008, *47*, 2674–2687.
275. Leyssens, T.; Peeters, D.; Harvey, J. N. Origin of enantioselective hydrogenation of ketones by RuH₂(diphosphine)(diamine) catalysts: A theoretical study. *Organometallics* 2008, *27*, 1514–1523.
276. French, S. A.; Di, T. D.; Zanotti-Gerosa, A.; Hancock, F.; Catlow, C. R. A. New insights into the enantioselectivity in the hydrogenation of prochiral ketones. *Chem. Commun.* 2007, 2381–2383.
277. Hadzovic, A.; Song, D.; MacLaughlin, C. M.; Morris, R. H. A mechanism displaying autocatalysis: The hydrogenation of acetophenone catalyzed by RuH(S-binap)(app) where app is the amido ligand derived from 2-Amino-2-(2-pyridyl)propane. *Organometallics* 2007, *26*, 5987–5999.
278. Di, T. D.; French, S. A.; Catlow, C. R. A. The H₂-hydrogenation of ketones catalyzed by ruthenium(II) complexes: A density functional theory study. *J. Mol. Struct. Theochem.* 2007, *812*, 39–49.
279. Hedberg, C.; Kaellstroem, K.; Arvidsson, P. I.; Brandt, P.; Andersson, P. G. Mechanistic insights into the phosphine-free RuCp-diamine-catalyzed hydrogenation of aryl ketones: Experimental and theoretical evidence for an alcohol-mediated dihydrogen activation. *J. Am. Chem. Soc.* 2005, *127*, 15083–15090.
280. Václavík, J. I.; Kuzma, M.; Přech, J.; Kačer, P. Asymmetric transfer hydrogenation of imines and ketones using chiral RuIICl(η6-p-cymene)[(S, S)-N-TsDPEN] as a catalyst: A computational study. *Organometallics* 2011, *30*, 4822–4829.
281. Handgraaf, J. W.; Reek, J. N. H.; Meijer, E. J. Iridium(I) versus ruthenium(II). A computational study of the transition metal catalyzed transfer hydrogenation of ketones. *Organometallics* 2003, *22*, 3150–3157.
282. Yamakawa, M.; Yamada, I.; Noyori, R. CH/pi attraction: The origin of enantioselectivity in transfer hydrogenation of aromatic carbonyl compounds catalyzed by chiral eta(6)-arene-ruthenium(II) complexes. *Angew. Chem. Int. Ed.* 2001, *40*, 2818–2821.
283. Noyori, R.; Yamakawa, M.; Hashiguchi, S. Metal-ligand bifunctional catalysis: A nonclassical mechanism for asymmetric hydrogen transfer between alcohols and carbonyl compounds. *J. Org. Chem.* 2001, *66*, 7931–7944.
284. Yamakawa, M.; Ito, H.; Noyori, R. The metal-ligand bifunctional catalysis: A theoretical study on the ruthenium(II)-catalyzed hydrogen transfer between alcohols and carbonyl compounds. *J. Am. Chem. Soc.* 2000, *122*, 1466–1478.

285. Petra, D. G. I.; Reek, J. N. H.; Handgraaf, J.-W.; Meijer, E. J.; Dierkes, P.; Kamer, P. C. J.; Brussee, J.; Schoemaker, H. E.; Van, L. P. W. N. M. Chiral induction effects in ruthenium(II) amino alcohol catalyzed asymmetric transfer hydrogenation of ketones: An experimental and theoretical approach. *Chem.* 2000, *6*, 2818–2829.

286. Alonso, D. A.; Brandt, P.; Nordin, S. J. M.; Andersson, P. G. Ru(arene)(amino alcohol)-catalyzed transfer hydrogenation of ketones: Mechanism and origin of enantioselectivity. *J. Am. Chem. Soc.* 1999, *121*, 9580–9588.

287. John, J. M.; Takebayashi, S.; Dabral, N.; Miskolzie, M.; Bergens, S. H. Base-catalyzed bifunctional addition to amides and imides at low temperature. A new pathway for carbonyl hydrogenation. *J. Am. Chem. Soc.* 2013, *135*, 8578–8584.

288. Hasanayn, F.; Morris, R. H. Symmetry aspects of H_2 splitting by five-coordinate d6 ruthenium amides, and calculations on acetophenone hydrogenation, ruthenium alkoxide formation, and subsequent hydrogenolysis in a model trans-Ru(H)2(diamine)(diphosphine) system. *Inorg. Chem.* 2012, *51*, 10808–10818.

289. Takebayashi, S.; Dabral, N.; Miskolzie, M.; Bergens, S. H. Experimental investigations of a partial Ru–O bond during the metal–ligand bifunctional addition in Noyori-type enantioselective ketone hydrogenation. *J. Am. Chem. Soc.* 2011, *133*, 9666–9669.

290. Hamilton, R. J.; Bergens, S. H. Direct observations of the metal-ligand bifunctional addition step in an enantioselective ketone hydrogenation. *J. Am. Chem. Soc.* 2008, *130*, 11979–11987.

291. Hamilton, R. J.; Bergens, S. H. An unexpected possible role of base in asymmetric catalytic hydrogenations of ketones. Synthesis and characterization of several key catalytic intermediates. *J. Am. Chem. Soc.* 2006, *128*, 13700–13701.

292. Hamilton, R. J.; Leong, C. G.; Bigam, G.; Miskolzie, M.; Bergens, S. H. A ruthenium–dihydrogen putative intermediate in ketone hydrogenation. *J. Am. Chem. Soc.* 2005, *127*, 4152–4153.

293. Hartmann, R.; Chen, P. Noyori's hydrogenation catalyst needs a Lewis acid cocatalyst for high activity. *Angew. Chem. Int. Ed. Engl.* 2001, *40*, 3581–3585.

294. Zhu, M. Effect of NH acidity on transfer hydrogenation of Noyori–Ikariya catalyst. *Catal. Lett.* 2016, 1–5.

295. Heusser, C. A. Efforts towards the cross coupling of acylsilanes and electrophiles via a metal-catalyzed Brook rearrangement, M.S. thesis. Boston College, 2013.

296. Pavlova, A.; Meijer, E. J. Understanding the role of water in aqueous ruthenium-catalyzed transfer hydrogenation of ketones. *ChemPhysChem* 2012, *13*, 3492–3496.

297. Moasser, B. *Aqueous Metal-Ligand Bifunctional Hydrogenation of Carbon Dioxide*, 2010.

298. Handgraaf, J.-W.; Meijer, E. J. Realistic modeling of ruthenium-catalyzed transfer hydrogenation. *J. Am. Chem. Soc.* 2007, *129*, 3099–3103.

299. Tanis, S. P.; Evans, B. R.; Nieman, J. A.; Parker, T. T.; Taylor, W. D.; Heasley, S. E.; Herrinton, P. M.; Perrault, W. R.; Hohler, R. A.; Dolak, L. A.; Hester, M. R.; Seest, E. P. Solvent and in situ catalyst preparation impacts upon Noyori reductions of aryl-chloromethyl ketones: Application to syntheses of chiral 2-amino-1-aryl-ethanols. *Tetrahedron: Asymmetry* 2006, *17*, 2154–2182.

300. Koike, T.; Ikariya, T. Mechanistic aspects of formation of chiral ruthenium hydride complexes from 16-electron ruthenium amide complexes and formic acid: Facile reversible decarboxylation and carboxylation. *Adv. Synth. Catal.* 2004, *346*, 37–41.

301. Macchioni, A. Ion pairing in transition-metal organometallic chemistry. *Chem. Rev.* 2005, *105*, 2039–2074.

302. Borissova, A. O.; Antipin, M. Y.; Perekalin, D. S.; Lyssenko, K. A. Crucial role of RuH interactions in the crystal packing of ruthenocene and its derivatives. *CrystEngComm* 2008, *10*, 827–832.

303. Belkova, N. V.; Shubina, E. S.; Epstein, L. M. Diverse world of unconventional hydrogen bonds. *Acc. Chem. Res.* 2005, *38*, 624–631.

304. Braga, D.; Grepioni, F.; Tedesco, E.; Biradha, K.; Desiraju, G. R. Hydrogen bonding in organometallic crystals. 6. X–H—M hydrogen bonds and M—(H–X) pseudo-agostic bonds. *Organometallics* 1997, *16*, 1846–1856.

305. Abdur-Rashid, K.; Faatz, M.; Lough, A. J.; Morris, R. H. Catalytic cycle for the asymmetric hydrogenation of prochiral ketones to chiral alcohols: Direct hydride and proton transfer from chiral catalysts trans-Ru(H)2(diphosphine) (diamine) to ketones and direct addition of dihydrogen to the resulting hydridoamido complexes. *J. Am. Chem. Soc.* 2001, *123*, 7473–7474.

306. Koike, T.; Ikariya, T. Reaction of 16-electron ruthenium and iridium amide complexes with acidic alcohols: Intramolecular C-H bond activation and the isolation of cyclometalated complexes. *Organometallics* 2005, *24*, 724–730.

307. Otsuka, T.; Ishii, A.; Dub, P. A.; Ikariya, T. Practical selective hydrogenation of α-fluorinated esters with bifunctional pincer-type ruthenium(II) catalysts leading to fluorinated alcohols or fluoral hemiacetals. *J. Am. Chem. Soc.* 2013, *135*, 9600–9603.

308. Schlaf, M.; Lough, A. J.; Maltby, P. A.; Morris, R. H. Synthesis, structure, and properties of the stable and highly acidic dihydrogen complex trans-[Os(η2-H$_2$)(CH$_3$CN)(dppe)2](BF$_4$)2. Perspectives on the influence of the trans ligand on the chemistry of the dihydrogen ligand. *Organometallics* 1996, *15*, 2270–2278.

309. Kubas, G. J. Activation of dihydrogen and coordination of molecular H2 on transition metals. *J. Organomet. Chem.* 2014, *751*, 33–49.

310. Gordon, J. C.; Kubas, G. J. Perspectives on how nature employs the principles of organometallic chemistry in dihydrogen activation in hydrogenases. *Organometallics* 2010, *29*, 4682–4701.

311. Kubas, G. J. Fundamentals of H$_2$ binding and reactivity on transition metals underlying hydrogenase function and H$_2$ production and storage. *Chem. Rev.* 2007, *107*, 4152–4205.

312. Chandrasekhar, S. The principle of microscopic reversibility in organic chemistry—A critique. *Res. Chem. Intermed.* 1992, *17*, 173–209.

313. Krupka, R. M.; Kaplan, H.; Laidler, K. J. Kinetic consequences of the principle of microscopic reversibility. *Trans. Faraday Soc.* 1966, *62*, 2754–2759.

314. Burwell, R. L.; Pearson, R. G. The principle of microscopic reversibility. *J. Phys. Chem.* 1966, *70*, 300–302.

315. Shubina, E.; Belkova, N.; Filippov, O.; Epstein, L. Weak interactions and M–H bond activation. In *Advances in Organometallic Chemistry and Catalysis*, John Wiley & Sons, Hoboken, NJ, 2013, pp. 97–109.

316. Filippov, O. A.; Belkova, N. V.; Epstein, L. M.; Lledós, A.; Shubina, E. S. Hydrogen-deuterium exchange in hydride chemistry: Dihydrogen bonded complexes as key intermediates. *Comput. Theor. Chem.* 2012, *998*, 129–140.

317. Belkova, N. V.; Epstein, L. M.; Shubina, E. S. Dihydrogen bonding, proton transfer and beyond: What we can learn from kinetics and thermodynamics. *Eur. J. Inorg. Chem.* 2010, *2010*, 3555–3565.

318. Kubas, G. J.; Heinekey, D. M. *Activation of Molecular Hydrogen*, John Wiley & Sons, 2010.

319. Kubas, G. J. *Dihydrogen and Other σ Bond Complexes*, Vol. 1, Elsevier, 2007.

320. Belkova, N. V.; Epstein, L. M.; Shubina, E. S. Peculiarities of hydrogen bonding and proton transfer equilibria of organic versus organometallic bases. *ARKIVOC* 2008, 120–138.

321. Karpfen, A. Cooperative effects in hydrogen bonding. In *Advances in Chemical Physics*, John Wiley & Sons, 2003, pp. 469–510.

322. Reichardt, C. Solvent effects on chemical-reactivity. *Pure Appl. Chem.* 1982, *54*, 1867–1884.

323. Kosower, E. M. *Introduction to Physical Organic Chemistry*, Wiley, 1968.

324. From this perspective, Bullock's ionic hydrogenation can be categorized as a H+/H– outer-sphere hydrogenation mechanism, see Ref. [325].

325. Bullock, R. M. Catalytic ionic hydrogenations. *Chem.* 2004, *10*, 2366–2374.

326. Ohkuma, T.; Doucet, H.; Pham, T.; Mikami, K.; Korenaga, T.; Terada, M.; Noyori, R. Asymmetric activation of racemic ruthenium(II) complexes for enantioselective hydrogenation. *J. Am. Chem. Soc.* 1998, *120*, 1086–1087.

327. Ohkuma, T.; Koizumi, M.; Doucet, H.; Pham, T.; Kozawa, M.; Murata, K.; Katayama, E.; Yokozawa, T.; Ikariya, T.; Noyori, R. Asymmetric hydrogenation of alkenyl, cyclopropyl, and aryl ketones. RuCl2(xylbinap)(1,2-diamine) as a precatalyst exhibiting a wide scope. *J. Am. Chem. Soc.* 1998, *120*, 13529–13530.

328. Ohkuma, T.; Koizumi, M.; Muñiz, K.; Hilt, G.; Kabuto, C.; Noyori, R. Trans-RuH(η1-BH4)(binap)(1,2-diamine): A catalyst for asymmetric hydrogenation of simple ketones under base-free conditions. *J. Am. Chem. Soc.* 2002, *124*, 6508–6509.

329. Berkessel, A.; Schubert, T. J. S.; Müller, T. N. Hydrogenation without a transition-metal catalyst: On the mechanism of the base-catalyzed hydrogenation of ketones. *J. Am. Chem. Soc.* 2002, *124*, 8693–8698.

330. Walling, C.; Bollyky, L. Homogeneous hydrogenation in the absence of transition-metal catalysts. *J. Am. Chem. Soc.* 1964, *86*, 3750–3752.

331. Walling, C.; Bollyky, L. Base-catalyzed homogeneous hydrogenation. *J. Am. Chem. Soc.* 1961, *83*, 2968–2969.

332. Chai, J.-D.; Head-Gordon, M. Long-range corrected hybrid density functionals with damped atom-atom dispersion corrections. *Phys. Chem. Chem. Phys.* 2008, *10*, 6615–6620.

333. Marenich, A. V.; Cramer, C. J.; Truhlar, D. G. Universal solvation model based on solute electron density and on a continuum model of the solvent defined by the bulk dielectric constant and atomic surface tensions. *J. Phys. Chem.* 2009, *113*, 6378–6396.

334. Matsuoka, A.; Sandoval, C. A.; Uchiyama, M.; Noyori, R.; Naka, H. Why p-Cymene? Conformational effect in asymmetric hydrogenation of aromatic ketones with a η6-arene/ruthenium(II) catalyst. *Chem.* 2015, *10*, 112–115.

335. Mooibroek, T. J.; Gamez, P.; Reedijk, J. Lone pair-[small pi] interactions: A new supramolecular bond? *CrystEngComm* 2008, *10*, 1501–1515.
336. Hopmann, K. H. Quantum chemical studies of asymmetric reactions: Historical aspects and recent examples. *Int. J. Quantum Chem.* 2015, *115*, 1232–1249.
337. Šterk, D.; Stephan, M.; Mohar, B. Highly enantioselective transfer hydrogenation of fluoroalkyl ketones. *Org. Lett.* 2006, *8*, 5935–5938.
338. Brandt, P.; Roth, P.; Andersson, P. G. Origin of enantioselectivity in the Ru(arene)(amino alcohol)-catalyzed transfer hydrogenation of ketones. *J. Org. Chem.* 2004, *69*, 4885–4890.
339. Battaglia, M. R.; Buckingham, A. D.; Williams, J. H. The electric quadrupole moments of benzene and hexafluorobenzene. *Chem. Phys. Lett.* 1981, *78*, 421–423.
340. Hiedrich, D. *The Reaction Path in Chemistry: Current Approaches and Perspectives.* Understanding Chemical Reaction Series 1995; 16, Kluwer, Dordrecht, the Netherlands, 1995.
341. Friedrich, A.; Drees, M.; Schmedt auf der Günne, J.; Schneider, S. Highly stereoselective proton/hydride exchange: Assistance of hydrogen bonding for the heterolytic splitting of H2. *J. Am. Chem. Soc.* 2009, *131*, 17552–17553.
342. Limbach, H. H.; Miguel Lopez, J.; Kohen, A. Arrhenius curves of hydrogen transfers: Tunnel effects, isotope effects and effects of pre-equilibria. *Phil. Trans. R. Soc. B* 2006, *361*, 1399–1415.
343. Miyazaki, T. Large isotope effects in tunneling chemical reactions. *J. Nucl. Sci. Tech.* 2002, *39*, 339–343.
344. Krishtalik, L. I. The mechanism of the proton transfer: An outline. *Biochim. Biophys. Acta* 2000, *1458*, 6–27.
345. Christov, S. G. Theory of proton transfer reactions in solution. *Chem. Phys.* 1992, *168*, 327–339.
346. Assuming no photochemistry involved.
347. Stanley, K.; Baird, M. C. Demonstration of controlled asymmetric induction in organoiron chemistry. Suggestions concerning the specification of chirality in pseudotetrahedral metal complexes containing polyhapto ligands. *J. Am. Chem. Soc.* 1975, *97*, 6598–6599.
348. Lecomte, C.; Dusausoy, Y.; Protas, J.; Tirouflet, J.; Dormond, A. Crystal structure and relative configuration of a titanocene complex having a planar chirality and a chirality centered on the titanium atom. *J. Organometal. Chem.* 1974, *73*, 67–76.
349. Espinet, P.; Casares, J. A. Conformational mobility in chelated square-planar Rh, Ir, Pd, and Pt complexes. *Phys. Organomet. Chem.* 2004, *4*, 131–161.
350. Espinet, P.; Casares, J. A. Conformational mobility in chelated square-planar Rh, Ir, Pd, and Pt complexes. In *Fluxional Organometallic and Coordination Compounds*, Vol. 4, Gielen, M.; Willem, R.; Wrackmeyer, B.; Eds.; Wiley & Sons, 2004, pp. 131–158.
351. Letko, C. S.; Heiden, Z. M.; Rauchfuss, T. B.; Wilson, S. R. Coordination chemistry of the soft chiral Lewis acid [Cp*Ir(TsDPEN)]+. *Inorg. Chem.* 2011, *50*, 5558–5566.
352. Soni, R.; Cheung, F. K.; Clarkson, G. C.; Martins, J. E. D.; Graham, M. A.; Wills, M. The importance of the N-H bond in Ru/TsDPEN complexes for asymmetric transfer hydrogenation of ketones and imines. *Org. Biomol. Chem.* 2011, *9*, 3290–3294.

353. Crabtree, R. H.; Uriarte, R. J. Electronic effects on the orientation of hydrogen addition to an iridium(I) complex. *Inorg. Chem.* 1983, *22*, 4152–4154.
354. Crabtree, R. H.; Demou, P. C.; Eden, D.; Mihelcic, J. M.; Parnell, C. A.; Quirk, J. M.; Morris, G. E. Dihydrido olefin and solvento complexes of iridium and the mechanisms of olefin hydrogenation and alkane dehydrogenation. *J. Am. Chem. Soc.* 1982, *104*, 6994–7001.
355 Crabtree, R. H.; Felkin, H.; Morris, G. E. Cationic iridium diolefin complexes as alkene hydrogenation catalysts and the isolation of some related hydrido complexes. *J. Organomet. Chem.* 1977, *141*, 205–215.
356. Lightfoot, A.; Schnider, P.; Pfaltz, A. Enantioselective hydrogenation of olefins with iridium–phosphanodihydrooxazole catalysts. *Angew. Chem. Int. Ed.* 1998, *37*, 2897–2899.
357. Pfaltz, A.; Blankenstein, J.; Hilgraf, R.; Hörmann, E.; McIntyre, S.; Menges, F.; Schönleber, M.; Smidt, S. P.; Wüstenberg, B.; Zimmermann, N. Iridium-catalyzed enantioselective hydrogenation of olefins. *Adv. Synth. Catal.* 2003, *345*, 33–43.
358. Cui, X.; Burgess, K. Catalytic homogeneous asymmetric hydrogenations of largely unfunctionalized alkenes. *Chem. Rev.* 2005, *105*, 3272–3296.
359. Källström, K.; Hedberg, C.; Brandt, P.; Bayer, A.; Andersson, P. G. Rationally designed ligands for asymmetric iridium-catalyzed hydrogenation of olefins. *J. Am. Chem. Soc.* 2004, *126*, 14308–14309.
360. Kallstrom, K.; Munslow, I.; Andersson, P. G. Ir-catalysed asymmetric hydrogenation: Ligands, substrates and mechanism. *Chem.* 2006, *12*, 3194–3200.
361. Roseblade, S. J.; Pfaltz, A. Iridium-catalyzed asymmetric hydrogenation of olefins. *Acc. Chem. Res.* 2007, *40*, 1402–1411.
362. Roseblade, S. J.; Pfaltz, A. Recent advances in iridium-catalysed asymmetric hydrogenation of unfunctionalised olefins. *Comptes Rendus Chimie* 2007, *10*, 178–187.
363. Li, S.; Zhu, S.-F.; Xie, J.-H.; Song, S.; Zhang, C.-M.; Zhou, Q.-L. Enantioselective hydrogenation of α-aryloxy and α-alkoxy α,β-unsaturated carboxylic acids catalyzed by chiral spiro iridium/phosphino-oxazoline complexes. *J. Am. Chem. Soc.* 2010, *132*, 1172–1179.
364. Cadu, A.; Andersson, P. G. Development of iridium-catalyzed asymmetric hydrogenation: New catalysts, new substrate scope. *J. Organomet. Chem.* 2012, *714*, 3–11.
365. Zhou, T. G.; Peters, B.; Maldonado, M. F.; Govender, T.; Andersson, P. G. Enantioselective synthesis of chiral sulfones by Ir-catalyzed asymmetric hydrogenation: A facile approach to the preparation of chiral allylic and homoallylic compounds. *J. Am. Chem. Soc.* 2012, *134*, 13592–13595.
366. Peters, B. K.; Zhou, T. G.; Rujirawanich, J.; Cadu, A.; Singh, T.; Rabten, W.; Kerdphon, S.; Andersson, P. G. An enantioselective approach to the preparation of chiral sulfones by Ir-catalyzed asymmetric hydrogenation. *J. Am. Chem. Soc.* 2014, *136*, 16557–16562.
367. Spindler, F.; Pugin, B.; Blaser, H.-U. Novel diphosphinoiridium catalysts for the enantioselective hydrogenation of N-arylketimines. *Angew. Chem. Int. Ed. Engl.* 1990, *29*, 558–559.
368. Ng Cheong Chan, Y.; Osborn, J. A. Iridium(III) hydride complexes for the catalytic enantioselective hydrogenation of imines. *J. Am. Chem. Soc.* 1990, *112*, 9400–9401.

369. Morimoto, T.; Achiwa, K. An improved diphosphine-iridium(I) catalyst system for the asymmetric hydrogenation of cyclic imines: Phthalimide as an efficient co-catalyst. *Tetrahedron: Asymmetry* 1995, *6*, 2661–2664.

370. Sablong, R.; Osborn, J. A. Asymmetric hydrogenation of imines catalysed by carboxylato(diphosphine)iridium(III) complexes. *Tetrahedron: Asymmetry* 1996, *7*, 3059–3062.

371. Zhu, G.; Zhang, X. Additive effects in Ir–BICP catalyzed asymmetric hydrogenation of imines. *Tetrahedron: Asymmetry* 1998, *9*, 2415–2418.

372. Hou, G.-H.; Xie, J.-H.; Yan, P.-C.; Zhou, Q.-L. Iridium-catalyzed asymmetric hydrogenation of cyclic enamines. *J. Am. Chem. Soc.* 2009, *131*, 1366–1367.

373. Ye, Z.-S.; Chen, M.-W.; Chen, Q.-A.; Shi, L.; Duan, Y.; Zhou, Y.-G. Iridium-catalyzed asymmetric hydrogenation of pyridinium salts. *Angew. Chem. Int. Ed.* 2012, *51*, 10181–10184.

374. Iimuro, A.; Yamaji, K.; Kandula, S.; Nagano, T.; Kita, Y.; Mashima, K. Asymmetric hydrogenation of isoquinolinium salts catalyzed by chiral iridium complexes: Direct synthesis for optically active 1,2,3,4-tetrahydroisoquinolines. *Angew. Chem. Int. Ed.* 2013, *52*, 2046–2050.

375. Ye, Z.-S.; Guo, R.-N.; Cai, X.-F.; Chen, M.-W.; Shi, L.; Zhou, Y.-G. Enantioselective iridium-catalyzed hydrogenation of 1- and 3-substituted isoquinolinium salts. *Angew. Chem. Int. Ed.* 2013, *52*, 3685–3689.

376. Zhao, D.; Glorius, F. Enantioselective hydrogenation of isoquinolines. *Angew. Chem. Int. Ed.* 2013, *52*, 9616–9618.

377. Rageot, D.; Woodmansee, D. H.; Pugin, B.; Pfaltz, A. Proline-based P, O ligand/iridium complexes as highly selective catalysts: Asymmetric hydrogenation of trisubstituted alkenes. *Angew. Chem. Int. Ed.* 2011, *50*, 9598–9601.

378. Liu, Y.; Gridnev, I. D.; Zhang, W. Mechanism of the asymmetric hydrogenation of exocyclic α,β-unsaturated carbonyl compounds with an iridium/BiphPhox catalyst: NMR and DFT studies. *Angew. Chem. Int. Ed.* 2014, *53*, 1901–1905.

379. Xie, J.-H.; Liu, X.-Y.; Xie, J.-B.; Wang, L.-X.; Zhou, Q.-L. An additional coordination group leads to extremely efficient chiral iridium catalysts for asymmetric hydrogenation of ketones. *Angew. Chem. Int. Ed.* 2011, *50*, 7329–7332.

380. Xie, J.-H.; Liu, X.-Y.; Yang, X.-H.; Xie, J.-B.; Wang, L.-X.; Zhou, Q.-L. Chiral iridium catalysts bearing spiro pyridine-aminophosphine ligands enable highly efficient asymmetric hydrogenation of β-aryl β-ketoesters. *Angew. Chem. Int. Ed.* 2012, *51*, 201–203.

381. Yang, X.-H.; Xie, J.-H.; Liu, W.-P.; Zhou, Q.-L. Catalytic asymmetric hydrogenation of δ-ketoesters: Highly efficient approach to chiral 1,5-diols. *Angew. Chem. Int. Ed.* 2013, *52*, 7833–7836.

382. Smidt, S. P.; Pfaltz, A.; Martínez-Viviente, E.; Pregosin, P. S.; Albinati, A. X-ray and NOE studies on trinuclear iridium hydride phosphino oxazoline (PHOX) complexes. *Organometallics* 2003, *22*, 1000–1009.

383. Gruber, S.; Neuburger, M.; Pfaltz, A. Characterization and reactivity studies of dinuclear iridium hydride complexes prepared from iridium catalysts with N, P and C, N ligands under hydrogenation conditions. *Organometallics* 2013, *32*, 4702–4711.

384. Mazet, C.; Smidt, S. P.; Meuwly, M.; Pfaltz, A. A combined experimental and computational study of dihydrido(phosphinooxazoline)iridium complexes. *J. Am. Chem. Soc.* 2004, *126*, 14176–14181.

385. Gruber, S.; Pfaltz, A. Asymmetric hydrogenation with iridium C, N and N, P ligand complexes: Characterization of dihydride intermediates with a coordinated alkene. *Angew. Chem. Int. Ed.* 2014, *53*, 1896–1900.
386. Brandt, P.; Hedberg, C.; Andersson, P. G. New mechanistic insights into the iridium–phosphanooxazoline-catalyzed hydrogenation of unfunctionalized olefins: A DFT and kinetic study. *Chem.* 2003, *9*, 339–347.
387. Cui, X.; Fan, Y.; Hall, M. B.; Burgess, K. Mechanistic insights into iridium-catalyzed asymmetric hydrogenation of dienes. *Chem.* 2005, *11*, 6859–6868.
388. Church, T. L.; Andersson, P. G. Iridium catalysts for the asymmetric hydrogenation of olefins with nontraditional functional substituents. *Coordin. Chem. Rev.* 2008, *252*, 513–531.
389. Mazuela, J.; Norrby, P. O.; Andersson, P. G.; Pamies, O.; Dieguez, M. Pyranoside phosphite-oxazoline ligands for the highly versatile and enantioselective Ir-catalyzed hydrogenation of minimally functionalized olefins. A combined theoretical and experimental study. *J. Am. Chem. Soc.* 2011, *133*, 13634–13645.
390. Church, T. L. R., T.; Andersson, P. G. Enantioselectivity in the iridium-catalyzed hydrogenation of unfunctionalized olefins. *Organometallics* 2010, *29*, 6769–6781.
391. Hopmann, K. H.; Bayer, A. On the mechanism of iridium-catalyzed asymmetric hydrogenation of imines and alkenes: A theoretical study. *Organometallics* 2011, *30*, 2483–2497.
392. Margalef, J.; Caldentey, X.; Karlsson, E. A.; Coll, M.; Mazuela, J.; Pamies, O.; Dieguez, M.; Pericas, M. A. A theoretically-guided optimization of a new family of modular P, S-ligands for iridium-catalyzed hydrogenation of minimally functionalized olefins. *Chem.* 2014, *20*, 12201–12214.
393. Landaeta, V. R.; Muñoz, B. K.; Peruzzini, M.; Herrera, V.; Bianchini, C.; Sánchez-Delgado, R. A. Imine hydrogenation by tribenzylphosphine rhodium and iridium complexes. *Organometallics* 2005, *25*, 403–409.
394. Fabrello, A.; Bachelier, A.; Urrutigoïty, M.; Kalck, P. Mechanistic analysis of the transition metal-catalyzed hydrogenation of imines and functionalized enamines. *Coordin. Chem. Rev.* 2010, *254*, 273–287.
395. Martín, M.; Sola, E.; Tejero, S.; Andrés, J. L.; Oro, L. A. Mechanistic investigations of imine hydrogenation catalyzed by cationic iridium complexes. *Chem.* 2006, *12*, 4043–4056.
396. Martín, M.; Sola, E.; Tejero, S.; López, J. A.; Oro, L. A. Mechanistic investigations of imine hydrogenation catalyzed by dinuclear iridium complexes. *Chem.* 2006, *12*, 4057–4068.
397. Fattorusso, E.; Taglialatela-Scafati, O., Eds., *Modern Alkaloids*, Wiley-VCH, Weinheim, Germany, 2008, and references therein;
      Gueritte, F.; Fahy, J. In *Anticancer Agents from Natural Products*, Cragg, G. M., Kingstom, D. G. I., Newman, D. J., Eds.; CRC Press, Boca Raton, FL, 2005, p. 123.
398. Chen, Q.-A.; Ye, Z.-S.; Duan, Y.; Zhou, Y.-G. Homogeneous palladium-catalyzed asymmetric hydrogenation. *Chem. Soc. Rev.* 2013, *42*, 497–511.
399. Wang, D.-S.; Chen, Q.-A.; Lu, S.-M.; Zhou, Y.-G. Asymmetric hydrogenation of heteroarenes and arenes. *Chem. Rev.* 2012, *112*, 2557–2590.
400. Kuwano, R.; Kashiwabara, M.; Sato, K.; Ito, T.; Kaneda, K.; Ito, Y. Catalytic asymmetric hydrogenation of indoles using a rhodium complex with a chiral bisphosphine ligand PhTRAP. *Tetrahedron: Asymmetry* 2006, *17*, 521–535.

401. Kuwano, R.; Kaneda, K.; Ito, T.; Sato, K.; Kurokawa, T.; Ito, Y. Highly enantio-selective synthesis of chiral 3-substituted indolines by catalytic asymmetric hydrogenation of indoles. *Org. Lett.* 2004, *6*, 2213–2215.
402. Kuwano, R.; Sato, K.; Kurokawa, T.; Karube, D.; Ito, Y. Catalytic asymmetric hydrogenation of heteroaromatic compounds, indoles. *J. Am. Chem. Soc.* 2000, *122*, 7614–7615.
403. Kuwano, R.; Kashiwabara, M. Ruthenium-catalyzed asymmetric hydrogenation of N-Boc-indoles. *Org. Lett.* 2006, *8*, 2653–2655.
404. Baeza, A.; Pfaltz, A. Iridium-catalyzed asymmetric hydrogenation of N-protected indoles. *Chem. – Eur. J.* 2010, *16*, 2036–2039.
405. Wang, D.-S.; Chen, Q.-A.; Li, W.; Yu, C.-B.; Zhou, Y.-G.; Zhang, X. Pd-catalyzed asymmetric hydrogenation of unprotected indoles activated by Brønsted acids. *J. Am. Chem. Soc.* 2010, *132*, 8909–8911.
406. Duan, Y.; Li, L.; Chen, M.-W.; Yu, C.-B.; Fan, H.-J.; Zhou, Y.-G. Homogenous Pd-catalyzed asymmetric hydrogenation of unprotected indoles: Scope and mechanistic studies. *J. Am. Chem. Soc.* 2014, *136*, 7688–7700.

chapter two

Other enantioselective reactions catalyzed by transition metals

2.1 Enantioselective reactions catalyzed by bifunctional Ru and Ir complexes

2.1.1 Mechanism of the Michael addition catalyzed by bifunctional Ru catalysts

Bifunctional molecular catalysts bearing chiral N-sulfonylated 1,2-diamine ligands, Ru(Tsdpen)(η^6-arene) (**1**) and (TsDPEN: N-(p-toluenesulfonyl)-1,2-diphenylethylenediamine) were originally developed by Noyori and Ikariya et al. for the asymmetric transfer hydrogenation of ketones and imines.[1–3] Chiral bifunctional amido Ru complexes **1** are the catalysts, and chiral amine hydride metal complexes exhibiting a metal/NH synergetic effect are involved in the catalytic cycle as intermediates.[1–3] These amino complexes are bifunctional, that is, a substrate is bound using two active sites in the same molecule. Numerous highly enantioselective reductions utilizing these bifunctional catalysts have been developed.[4–13]

1a:	R = Me	R_n =	1,2,3,4,5,6-Me$_6$
1b:	Me		1,3,5-Me$_3$
1c:	Ts		1,3,5-Me$_3$
1d:	Ts		1,2,3,4,5,6-Me$_6$
1e:	2,4,5,6-Me$_5$C$_6$		1,2,3,4,5,6-Me$_6$
1f:	Ts		1-Me-4-Pr^i

Recently, the mechanism of enantioselective Michael addition of malonic esters and keto esters to cyclic conjugated enones catalyzed by Ru bifunctional complexes **1** was studied in detail.[14]

NMR investigation of the stoichiometric reactions of chiral amido Ru complexes, Ru(N-sulfonylated dpen)(η6-arene) **1a–c**, with dimethyl malonate **2** and β-keto ester **3** revealed that at lowered temperatures deprotonation proceeds in a stereoselective manner to provide corresponding amine complexes **4, 5** and **6, 7** (Scheme 2.1).

In the case of the malonate **2**, the C-bound complexes **4a,c** formed almost exclusively, whereas the reaction with the β-keto ester **3** gave complicated equilibria of numerous species. The computations showed that

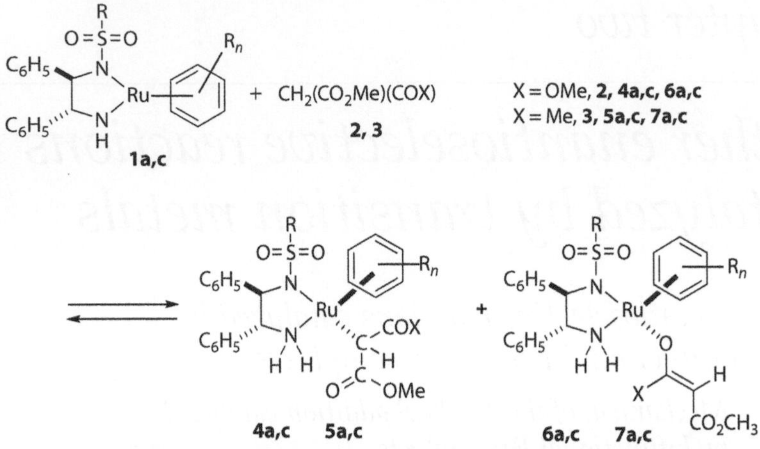

Scheme 2.1 Formation of amino complexes 4–7.

either of these species could be kinetically available for further reaction with enone, hence all possibilities were accurately investigated.

Further computational studies showed that although the C-bound complexes **4, 5** are incapable of activating a substrate, the O-bound complexes **6, 7** contain a 3D cavity suitable for accommodating a molecule of cyclopentenone (**8**). However, the computed pathways for enantioselective C–C bond formation starting from **6, 7** were characterized by unreasonably high activation barriers.[14]

Hence, further possibilities were investigated. It has been found that the chelate ionic pair, for example **9**, which can easily form in the equilibrium mixture of the amine complexes, is capable of coordinating cyclopentenone **8** to effectuate the C–C bond formation with a low activation barrier (Scheme 2.2).

Figure 2.1 displays the optimized structures of the transition states of the reaction steps important for understanding the origin of enantioselectivity in this reaction. Two transition states shown in Figure 2.1 (top) differ by the mode of coordination of the molecule of **8** to the Ru atom. The transition state leading to the formation of the *S*-enantiomer of the product (Figure 2.1, top, right) is significantly less stable than that leading to the *R*-enantiomer (Figure 2.1, top, left) due to the unavoidable interaction between the CH_2CH_2 moiety of the substrate and the coordinated arene of the catalyst. The free energy gap for 8.2 kcal/mol could result in perfect enantioselectivity. However, the ion pair **9** can easily invert configuration of its Ru atom making another side of the chelate cycle available for the bifunctional binding of the substrate. In that case the bifunctional coordination of cyclopentenone **8** resulting in the formation of *S*-isomer leads to

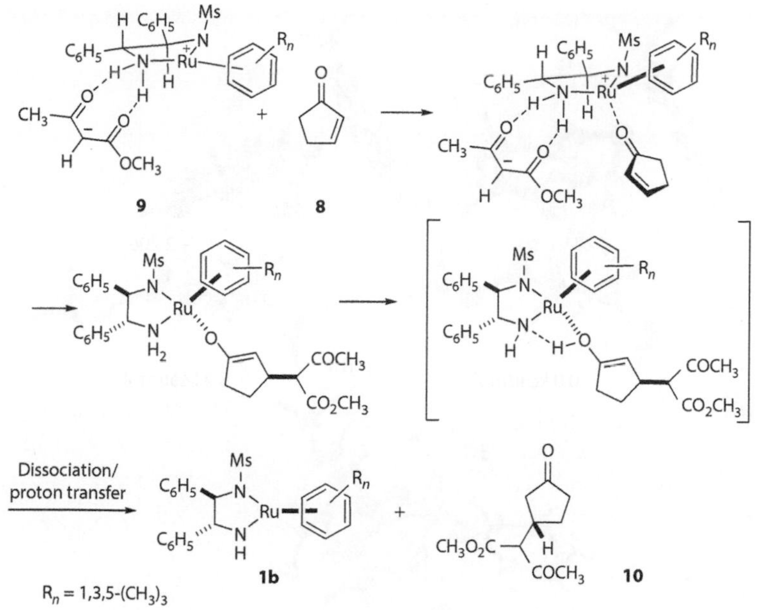

Scheme 2.2 Mechanism of Michael addition catalyzed by Ru complexes **1a,c**. (Data from Gridnev, I. D. et al., *J. Am. Chem. Soc.*, 132, 16637–16650, 2010.)

the transition state (Figure 2.1 below) that is only 1.8 kcal/mol less stable than the most stable transition state responsible for the formation of the *R*-product.

Hence, although the asymmetric environment of the Ru catalyst is efficient enough for discriminating the *R*- and *S*-pathways within a single mechanism almost perfectly, the interference of another possible way of the substrate activation reduces the overall optical yield of the catalytic reaction.[14]

Systematic study of a series of catalyst–substrate combinations revealed that the experimentally observed sense of enantioselection was consistently explained by computational analysis. The experimentally detected tendency of increasing ee with the bulk of the coordinated arene in Ru complex is reproduced computationally by changes in the difference of either zero point vibrational energy (ZPVE)-corrected energies or Gibbs free energies for the *S*- and *R*-pathways (Table 2.1).[14]

2.1.2   *Mechanism of C–C and C–N bond forming reactions catalyzed by bifunctional Ir catalysts*

Chiral bifunctional Ir complexes were recently reported to catalyze asymmetric direct amination of *tert*-butyl α-phenyl-α-cyanoacetates, for example **11** using dimethyl azodicarboxylates, for example **12** to

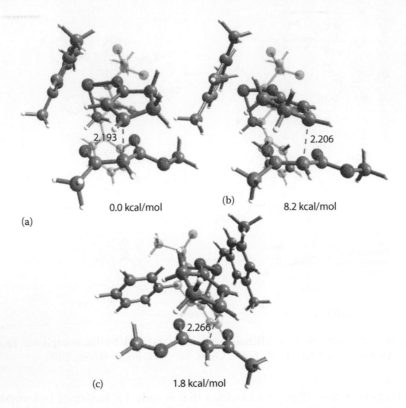

Figure 2.1 Transition states for the Michael addition reaction of malonic ester **61** and cyclopentenone **8** leading to **10**(*R*) (a) and to **10**(*S*) (b and c) and their computed relative energies. (Adapted with permission from Gridnev, I. D. et al., *J. Am. Chem. Soc.*, 132, 16637–16650. Copyright 2010 American Chemical Society.)

produce the corresponding hydrazine adduct **13** in 99% yield and high *R*-enantioselectivity (Scheme 2.3).[15] Interestingly, the use of the same catalyst **14** and the same phenyl-α-cyanoacetate **11** for the reaction with acetylenic ester **15** almost quantitatively gave the corresponding chiral adduct **16** with high *S*-enantioselectivity (Scheme 2.3).[16]

NMR investigation of the reaction between the Ir complex **14** and α-cyanoacetate **11** showed that a stereoselective deprotonation reaction takes place providing an additional chiral center on the transition metal atom and yielding N-bound amino complexes **17** and **18** (Scheme 2.4). Fast equilibration of the two species **17** and **18** was detected via NMR experiments at lowered temperatures.[17]

Computations of the possible reaction pathways showed that in the C–N bond forming reaction, the diazoester **12** is capable of undergoing a bifunctional activation by either of the catalyst–substrate complexes **17**, **18**(*si*) or **18**(*re*) to give the adducts **19**, **20**, or **21**, respectively (Scheme 2.5,

Table 2.1 Difference in energies (ZPVE [zero point vibrational energy] corrected) and free energies (kcal/mol) of the computed transition states for the formation of *S*- and *R*-products

Substrate	Sulfonyl group of the catalyst	Coordinated arene of the catalyst	ΔE_{ZPVE}	ΔG (298)	ΔG_{exp}[a]
2	Ms	Hexamethylbenzene	4.7	6.3	
		Pentamethylbenzene	3.7	5.3	
		Mesitylene	1.7	2.6	
		p-Cymene	0.3	1.8	
	Ts	Hexamethylbenzene	3.7	5.1	2.3
		Pentamethylbenzene	2.6	4.0	2.1
		Mesitylene	1.1	3.0	1.2
		p-Cymene	0.6	1.8	0.9
3	Ms	Hexamethylbenzene	4.3	5.8	
		Pentamethylbenzene	4.0	6.7	
		Mesitylene	1.8	3.6	
		p-Cymene	0.8	2.0	
	Ts	Hexamethylbenzene	3.4	5.0	
		Pentamethylbenzene	2.5	3.6	
		Mesitylene	0.6	2.2	
		p-Cymene	0.8	2.0	

Source: From Ikariya, T. and Gridnev, I. D. *Top. Cat.*, *53*, 894–901, 2010; Gridnev, I. D. et al., *J. Am. Chem. Soc.*, 132, 16637–16650, 2010. With permission.

[a] Calculated from the ee's values observed in the corresponding catalytic reactions.

Figure 2.2). Further C–N bond formation is the stereoselective stage. The structurally similar **TS1(R)** and **TS(S)** are nicely discriminated by 5.2 kcal/mol due to the unavoidable hindrance between the carboxymethyl substituent in **12** and BOC substituent in **21** that is absent in **19** (Figure 2.2, Scheme 2.5). However, the structurally different mode of the substrate bifunctional activation available in the H-bound complex **17** and realized in the adduct **20** results in the possibility of the formation of the *S*-product via **TS2** which is only 2.2 kcal/mol higher in energy than the **TS1(R)** (Scheme 2.5, Figure 2.2).

Hence, similarly to the situation described in the previous section, the experimentally observed optical yields in this catalytic reaction are determined not via the competition of the structurally similar transition states **TS1(R)** and **TS1(S)** (which would result in a practically perfect enantioselection), but between the transition states **TS1(R)** and **TS2** that apply the different mechanism for the substrate activation. An important practical conclusion is that it is useless trying to improve the optical yield by further increasing the size of the substituent in α-cyanoacetate, since the BOC substituent in **70** is good enough to exclude the possibility

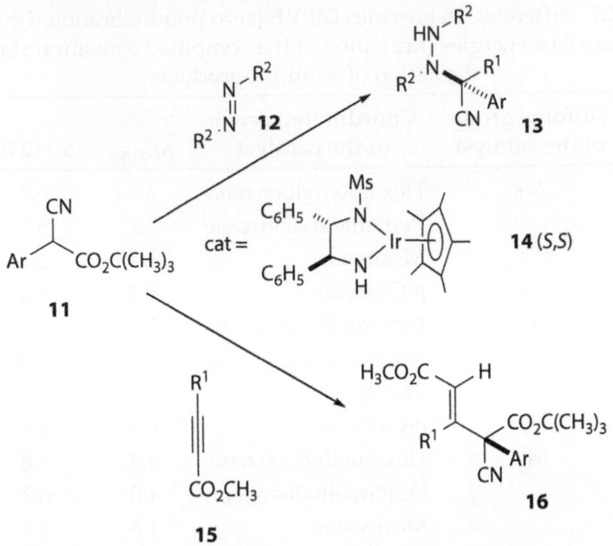

Scheme 2.3 Opposite stereochemical outcome of the C—N and C—C bond forming reactions using the same catalyst **14** and phenyl-α-cyanoacetate **11**.

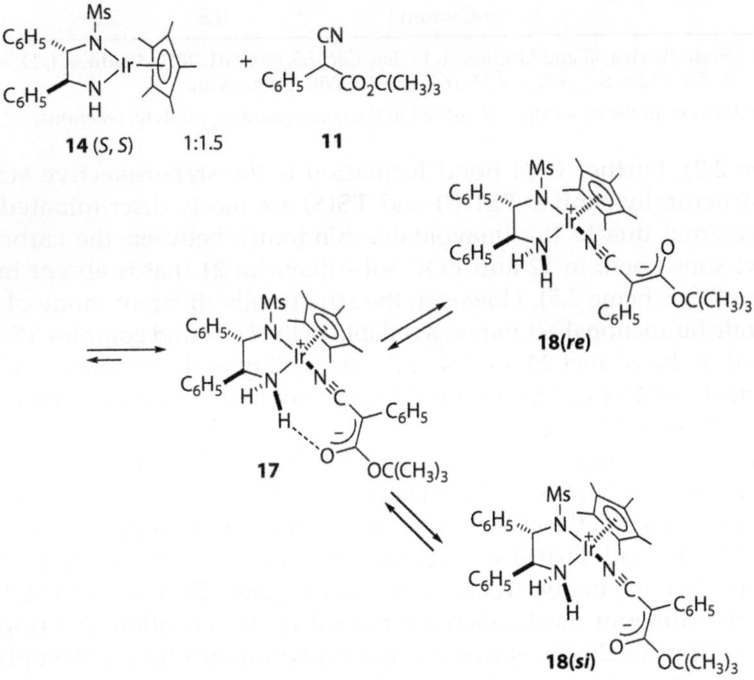

Scheme 2.4 Reaction of catalyst **14** with phenyl-α-cyanoacetate **11**.

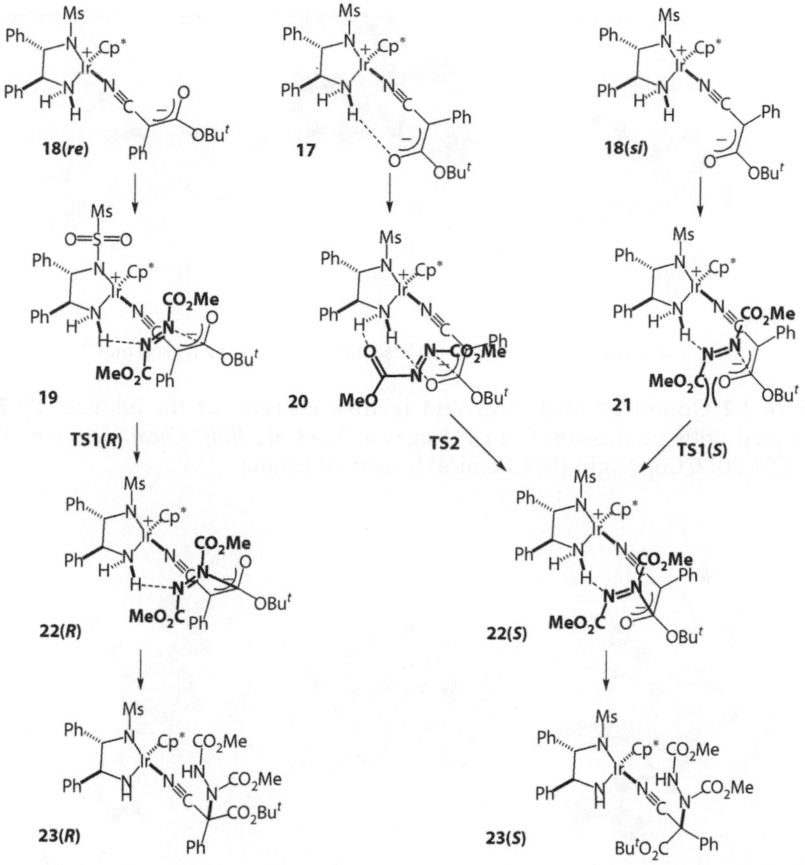

Scheme 2.5 Three competing pathways in the C—N bond forming reaction catalyzed by the Ir complex **14**. (Data from Hasegawa, Y. et al., *Bull. Chem. Soc. Jpn.*, 85, 316–334, 2012.)

of the reaction via **TS1(S)**. On the other hand, one can consider trying to affect the position and rates of the equilibria in the reaction pool, for example, by changing the temperature of the catalytic reaction. Thus, the best optical yield for the reaction of **11** and **12** catalyzed by **14** observed at –40°C (97% ee, corresponds to 2.1 kcal/mol difference in the stabilities of the competing transition states) stands well with the computed value of 2.2 kcal/mol difference in energies between **TS1(R)** and **TS2**, whereas at higher temperatures lower optical yields were observed (Figure 2.3).[17]

In view of the previous discussion, it is easy to understand why in the case of acetylenic esters the opposite sense of enantioselection is observed. Instead of the angular molecule of diazoester in complexes **19–21**, the linear molecule of acetylenic diester must be bifunctionally activated. As a result,

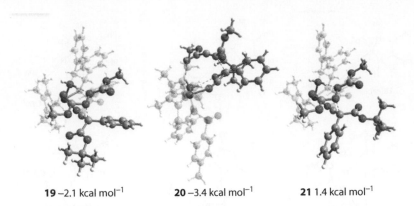

19 −2.1 kcal mol⁻¹   **20** −3.4 kcal mol⁻¹   **21** 1.4 kcal mol⁻¹

Figure **2.2** Optimized structures and relative energies of the adducts **19–21**. (Adapted with permission from Hasegawa, Y. et al., *Bull. Chem. Soc. Jpn.*, 85, 316–334, 2012. Copyright the Chemical Society of Japan.)

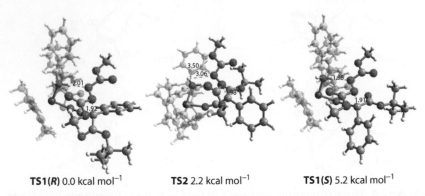

TS1(*R*) 0.0 kcal mol⁻¹   **TS2** 2.2 kcal mol⁻¹   **TS1(*S*)** 5.2 kcal mol⁻¹

Figure **2.3** Optimized structures and relative energies of the transition states **TS1(R)**, **TS2**, and **TS1(S)**. (Adapted with permission from Hasegawa, Y. et al., *Bull. Chem. Soc. Jpn.*, 85, 316–334, 2012. Copyright the Chemical Society of Japan.)

the stereochemical preferences in the similar adducts **14–16** (and in the corresponding transition states) become opposite (Scheme 2.6, Figure 2.4).

The 1,4-bifunctional activation of the substrate in **25** closely resembles that in **20**. However, due to the linear structure of the substrate **25**, the carboxymethyl substituents can effectively avoid the hindrance from the substituents of the coordinated α-cyanoacetate that makes **TS4** the lowest in energy transition state leading to the C–C bond formation. The 1,2-activation seen for the diazo ester in the adducts **19** and **21** (Scheme 2.4) is not possible in the case of the substrate **15**, the activation in 1,4 manner requires the movement of the substrate toward the bulky substituents of the α-cyanoacetate (Scheme 2.5). As a result,

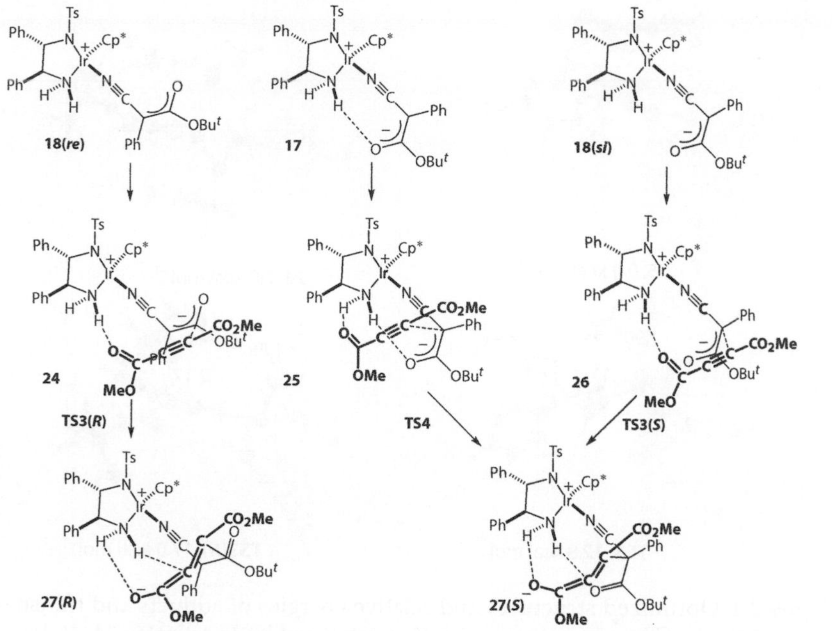

Scheme 2.6 Competing pathways in the C—C bond forming reaction catalyzed by the Ir complex **14**. (Adapted with permission from Hasegawa, Y. et al., *Bull. Chem. Soc. Jpn.*, 85, 316–334, 2012. Copyright the Chemical Society of Japan.)

the pathway involving **TS3(R)** that was computed to be 4.2 kcal/mol higher in energy than **TS4**, does not contribute to the flux of catalysis. Experimentally observed optical yields in the range 80%–95% ee have to be explained by the interference of the pathways involving monofunctional activation.[17]

The direct product of the C–C bond formation stage, intermediate **27(S)** has a bent allenic structure as shown in Schemes 2.6 and 2.7. The high Z/E ratios of the reaction product suggest that the proton transfer occurs directly from nitrogen to carbon, since the protonation of the oxygen atom would yield a hydroxyallene intermediate with equal potential for transferring the proton in two alternative positions (Scheme 2.7). However, before the proton transfer could take place, the negatively charged allenic unit must change its bending mode to **28(S)** yielding the intermediate **29(S)**. Computations show that this transformation as well as the following proton transfer is characterized by very low activation barriers of 1.6 and 1.9 kcal/mol, respectively. Hence, the high Z-selectivity of the major S-product is nicely explained by facile intramolecular proton transfer as shown in Scheme 2.7.

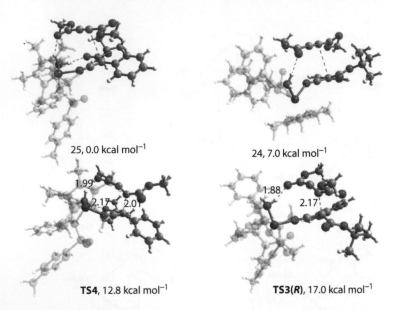

25, 0.0 kcal mol⁻¹   24, 7.0 kcal mol⁻¹

TS4, 12.8 kcal mol⁻¹   **TS3(R)**, 17.0 kcal mol⁻¹

Figure 2.4 Optimized structures and relative energies of adducts and transition states in the C–C bond formation reaction catalyzed by Ir complex **14**. (Adapted with permission from Hasegawa, Y. et al., *Bull. Chem. Soc. Jpn.*, 85, 316–334, 2012. Copyright the Chemical Society of Japan.)

On the other hand, although the rebending of the allenic unit is even more facile in the case of the intermediate **27(R)**, the proton transfer from the resulting **28(R)** was computed to have a much higher activation barrier of 10.5 kcal/mol (Figure 2.5, right). Taking into account the reversibility of the C–C bond formation step, this could serve as an additional stereo-discriminating factor favoring the formation of the *S* product. However, most probably this effect is leveled by the intermolecular protonation.

Much higher barrier of the intramolecular proton transfer in the case of **28(R)** compared to **28(S)** is explained by the relative instability of the corresponding transitions state **TS6(R)** compared to **TS6(S)** (Figure 2.5). The **TS6(S)** is very close in structure to the intermediate **27(S)**, the carboxymethyl substituent comfortably residing over the flat phenyl ring. On the other hand, in the transition state **TS6(R)** the necessary linear arrangement of the nitrogen, hydrogen, and carbon atoms can be only achieved at the expense of rotating the whole MeO₂C–C=C=C–CO₂Me unit for about 40° around the freshly formed C–C bond (Figure 2.5). Similarly to the C–C bond formation step, the main stereoregulating interaction is the hindrance from the BOC group of the cyanoester.

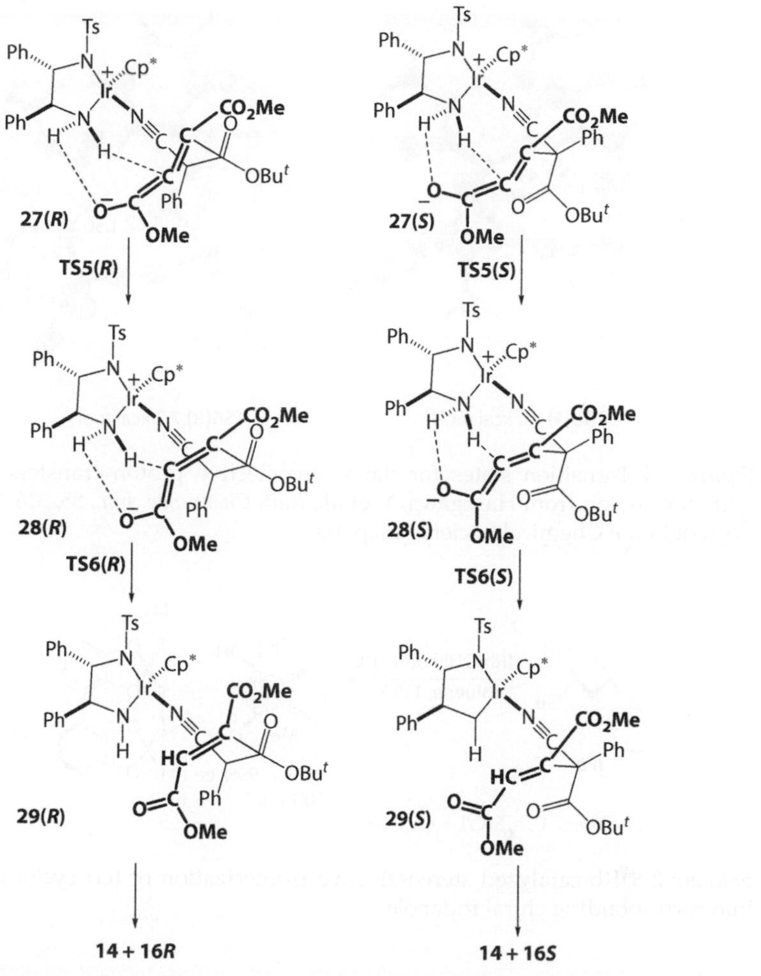

Scheme 2.7 Stereoselective proton transfer in bent allenic intermediates **27(R)** and **27(S)**. (Adapted with permission from Hasegawa, Y. et al., *Bull. Chem. Soc. Jpn.*, 85, 316–334, 2012. Copyright the Chemical Society of Japan.)

2.2 Rh-catalyzed stereoselective isomerization of tert-cyclobutanols into chiral indanoles

2.2.1 Overview

Activation of small, strained rings (e.g., cyclopropanes, cyclobutanes) may serve as a very interesting transformation to prepare various useful compounds such as α-tetralones,[18] indanoles,[19,20] and lactones[21,22] because ring opening releases ring strain and thus provides the driving force

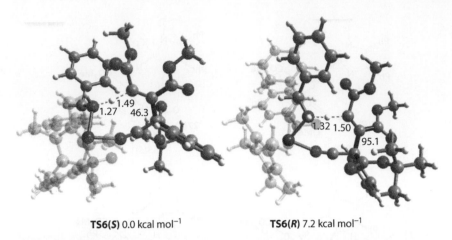

TS6(S) 0.0 kcal mol⁻¹   **TS6(R)** 7.2 kcal mol⁻¹

Figure 2.5 Transition states for the stereoselective proton transfer. (Adapted with permission from Hasegawa, Y. et al., *Bull. Chem. Soc. Jpn.*, 85, 316–334, 2012. Copyright the Chemical Society of Japan.)

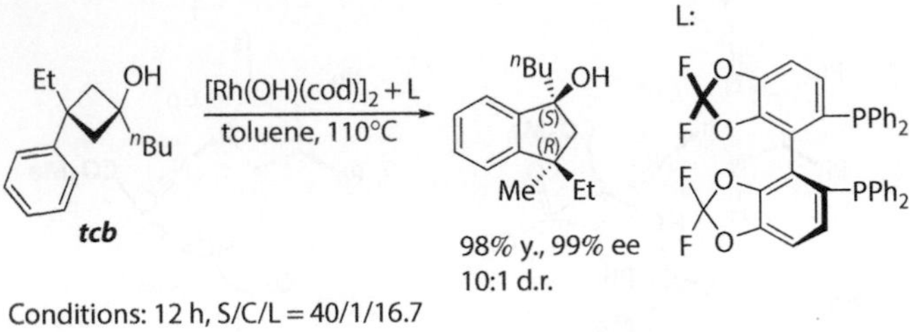

Scheme 2.8 Rh-catalyzed stereoselective isomerization of tert-cyclobutanol **tcb** into corresponding chiral indanole.

for C–C activation of the ring moiety. Cramer and coworkers reported Rh-catalyzed diastereoselective synthesis of indanoles with excellent ee up to 99%, although with relatively poor d.r. up to 20:1.[23] Representative example with *tert*-cyclobutanol *tcb* is shown in Scheme 2.8. Catalytic cycle, origin of enantio- and diastereoselectivity have been studies by Dang[24] and coworkers and will be summarized in this chapter.

2.2.2 *Catalytic cycle*

The proposed catalytic cycle (Cramer,[23] Dang[24]) is shown in Figure 2.6.[24] The active intermediate of the catalytic cycle, Rh[I] complex **A** is obtained by the reaction of the catalyst precursor [Rh(OH)(cod)]₂ (cod = cyclooctadiene) with added difluorophos ligand through monomerization and following cyclooctadiene dissociation, respectively. The catalytic cycle starts with

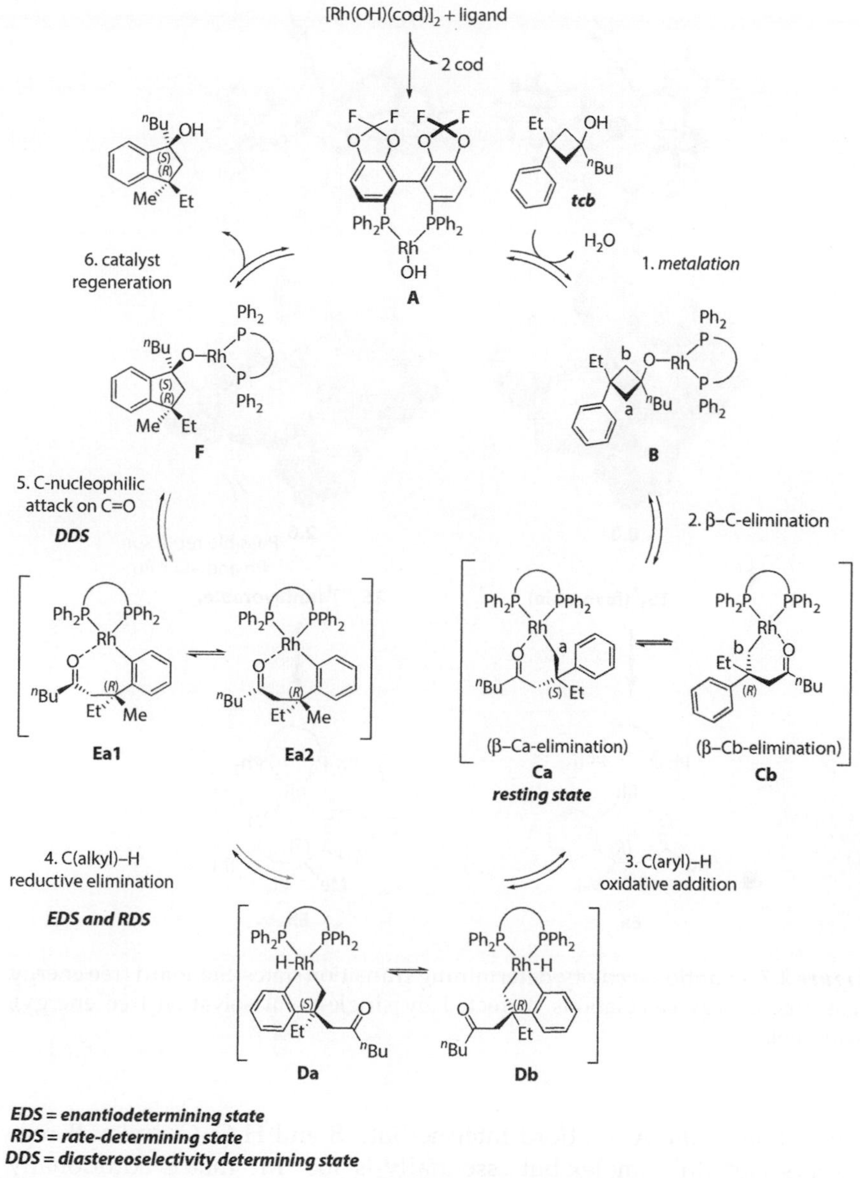

Figure 2.6 Computed catalytic cycle for Rh-Catalyzed Stereoselective Isomerization of *tert*-Cyclobutanol *tcb* under DFT/B3LYP/LANL2DZ(Rh)/6-31G*(all other) level of theory in the gas-phase. (Data from Yu, H. et al., *Chem. Eur. J.*, 20, 3839–3848, 2014.)

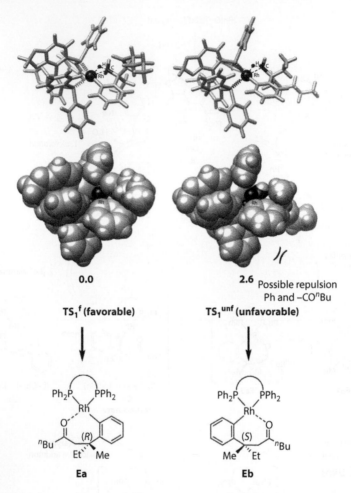

Figure 2.7 Enantioselectivity-determining transition states. Standard free energy (gas free energy calculations corrected by single-point solvation free energy), kcal/mol.

metalation of the **A** to afford intermediate **B** and H_2O. Complex **B** represents $14e^-$ Rh^I complex but essentially is $16e^-$ Rh^I that is additionally stabilized by one weak C–H sigma interaction of one methylene group of remaining substrate part with metal (four coordination positions, distorted square-planar geometry). Two isomers are possible for **B** (one stabilized by C–H sigma interaction of one methylene group of cyclobotanol ring, and one by one methylene group of ethyl), each affords complexes

Ca and **Cb** via β–C elimination of a or b-C atom of the cyclobutane ring, respectively, as shown in Figure 2.6. The diastereomeric Rh^I complexes **Ca** and **Cb** with terminally coordinated C=O oxygen atom exist in thermal equilibrium via intermediate **B**; **Ca** calculated to be the resting state of the catalytic cycle. Both **Ca** and **Cb** are converted into Rh^{III} complexes **Da** and **Db**, respectively, as a result of C(aryl)–H oxidative addition. Similarly to **Ca** and **Cb**, Rh^{III} complexes **Da** and **Db** exist in thermal equilibrium prior to irreversible rate-determining C(alkyl)–H reductive elimination states TS_1^f and TS_1^{unf}, respectively. This step proceeds effectively only from intermediate **Da** (27.8 kcal·mol^{-1} relative to the resting state, complex **Ca**) to afford Rh^I complex **Ea** with fixed absolute R-configuration on quaternary carbon atom (TS_1^f, see Figure 2.7). The opposite diastereomeric pathway that occurs from intermediate **Db** is uphill by 2.6 kcal·mol^{-1} on free energy scale (TS_1^{unf}, see Figure 2.7). This energy difference determines the observed enantioselectivity (calculated 97.7% ee at 25°C, *cf.* 99% ee from experiment at 110°C). The diastereoselectivity of the reaction is determined further by the energy difference between diastereomeric transition states corresponding to C-nucleophilic attack of C(arene) atom on carbonyl group. During this process, synchronous breaking of Rh–C(arene) and Rh←O=C and formation of C(O)–C and O–Rh bonds occur, respectively from two conformational isomers of **Ea** (**Ea1** and **Ea2** on Figure 2.6). The intermediates differ from each other in the conformation of the seven-membered ring, in which prochiral carbonyl group maybe C-attacked in two positions. The calculated energy difference of 1.7 kcal·mol^{-1} corresponds to the diastereoselectivity of 19:1, *cf.* d.r. 10:1 from experiment (Figure 2.8). Gas-phase calculations corrected by single-point solvation-free energy thus reliably reproduce both sense and order of enantio- and diastereoselection.

2.3 Os-catalyzed asymmetric dihydroxylation (Sharpless reaction)

2.3.1 Overview

Sharpless asymmetric dihydroxylation (AD) reaction or osmium-catalyzed dihydroxylation, Scheme 2.9, was initially developed in 1980s (1980: stoichiometric version,[25] 1988: catalytic version[26]) for the preparation of chiral diols from olefins.[27-31] The most popular standard set of reactants called as AD-mix-β or AD-mix-α has been developed for this reaction[32] and is still intensively used for producing the chiral products with up to 99% ee.[33,34] Commercially available AD-mix is composed of potassium osmate $K_2OsO_2(OH)_4$ and powdered $K_3Fe(CN)_6$ and K_2CO_3, respectively. β and

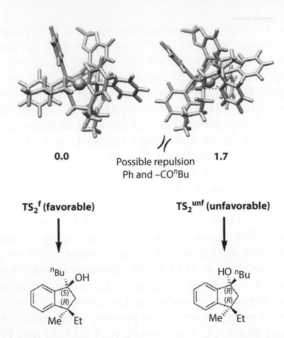

Figure 2.8 Diastereoselectivity-determining transition states. Standard free energy (gas free energy calculations corrected by single-point solvation free energy), kcal/mol.

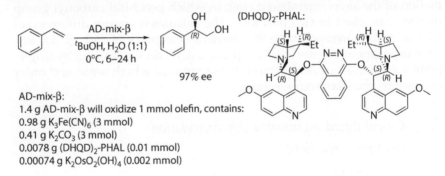

AD-mix-β:
1.4 g AD-mix-β will oxidize 1 mmol olefin, contains:
0.98 g $K_3Fe(CN)_6$ (3 mmol)
0.41 g K_2CO_3 (3 mmol)
0.0078 g $(DHQD)_2$-PHAL (0.01 mmol)
0.00074 g $K_2OsO_2(OH)_4$ (0.002 mmol)

Scheme 2.9 Classical sharpless asymmetric dihydroxylation reaction.

α denote the nature of the chiral ligand used. In the first case (β), ligand is abbreviated as $(DHQD)_2$-PHAL and represents phthalazine derivative of dihydroquinidine; in the second (α), ligand is called $(DHQ)_2$-PHAL and represents phthalazine derivative of dihydroquinine. Both dihydroquinidine and dihydroquinine are cinchona alkaloids[35] and are diastereomers due to the presence of the ethyl group at C3. Because of this fact, although ligands lead to diols of opposite configuration, the ee's are usually not identical. Each C_2-symmetric $(DHQD)_2$-PHAL or $(DHQ)_2$-PHAL has overall eight stereocenters as shown in Scheme 2.9 for $(DHQD)_2$-PHAL.

For styrene as a substrate, the AD-mix-β mixture gives the (R)-1-phenyl-1,2-ethandiol with 97% ee (S/C = 500) as shown in Scheme 2.9.[32]

2.3.2 Catalytic cycle

It was suggested that a stepwise [2+2] mechanism (involving a metallaoxetane intermediate)[36-38] or a concerted [3+2] pathway[39,40] may take place for this reaction. The latter, presently accepted,[41-45] is depicted in Figure 2.9. The reaction proceeds in two phases with $K_3Fe(CN)_6$ as the stoichiometric reoxidant.[46] In organic layer, chiral cinchona alkaloid derivative (ligand L) binds to OsO_4 via only one tertiary N-atom to afford 1:1 osmium(VIII)-ligand complex **A**. This reacts with olefin via a concerted multibond TS to afford osmium(VI) monoglycolate ester **B**. The later undergoes hydrolysis, releasing chiral diol and ligand to the organic layer and Os(VI) complex **C** to the aqueous layer before its reoxidation can occur to **D**, and consequently regeneration of **A** in organic layer.

2.3.3 Enantioselectivity

Os-catalyzed dihydroxylation of alkenes was probably one of the first density functional theory (DFT)-based investigations of a metal-catalyzed asymmetric reaction (Norrby, Sharpless).[36,47] The latest available computational work[35,48-52] devoted to the stereocontrol of this reaction was

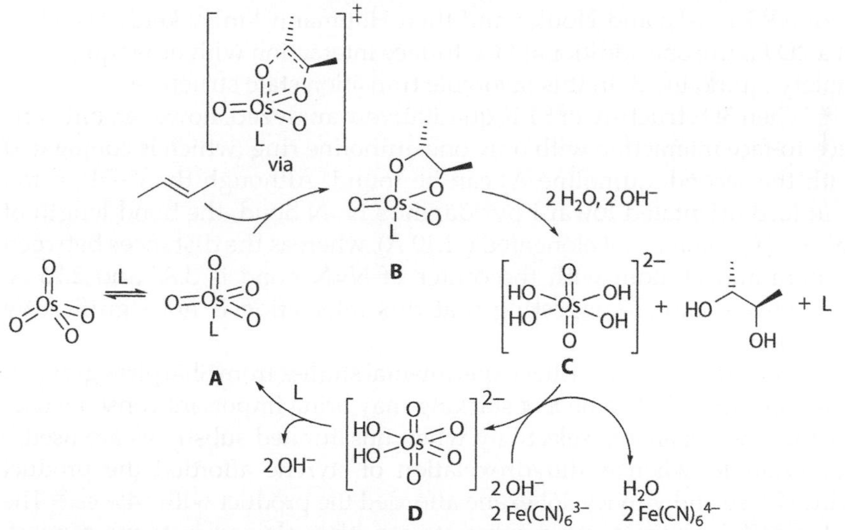

Figure 2.9 Catalytic cycle for the asymmetric dihydroxylation reaction in two phases $^tBuOH-H_2O$ with $K_3Fe(CN)_6$ as the stoichiometric reoxidant. The ligand is bound via tertiary nitrogen atom.

performed more than 15 years ago by Maseras and coworkers by means of hybrid QM/MM calculations with IMMOM method (gas-phase) and for styrene as a substrate.[53] Full model of the "second generation" ligand (DHQD)$_2$-PYDZ (bis(dihydroquinidine)-3,6-pyridazine) earlier introduced[54,55] by Corey was computed. The mono-OsO$_4$ complex of (DHQD)$_2$-PYDZ bounded via tertiary nitrogen atom was considered.[53] The dihydroxylation of styrene at 0°C using (DHQD)$_2$-PYDZ ligand afforded the final *R*-configurated product in 96% ee (*cf.* 97% ee obtained with (DHQD)$_2$-PHAL under the same conditions as shown in Scheme 2.9).[54] Twelve different possible reaction paths were defined from three regions of entry of the substrate via [3+2] concerted TS (three available O=Os=O) and its four possible orientations (position of Ph-substituent on "ethylene" with respect to two oxygen atoms).[53] Three lower energy saddle points correspond to the transition states leading to *R*-product (**S1**: 0.0 kcal·mol^{-1}; **S2**: 0.10 kcal·mol^{-1}) and *S*-product (**S3**: 2.65 kcal·mol^{-1}), respectively.[53] For each transition state structure, the substrate is located in the "pocket" of the catalyst–ligand complex (so-called B region[53,56]). The available structure corresponding to **S1** is shown in Figure 2.10.

Based on decomposition of the MM3 van der Waals interaction energy (IE) between substrate and catalyst it was concluded by Maseras and coworkers that two major factors stabilize **S1**: face-to-face π–π stacking interaction between substrate and one quinoline moiety (two conjugated aromatic rings, abbreviated as quinoline A) as well as face-to-edge interaction (C–H...N=N hydrogen bonding) between styrene and pyradizine.[53] Krenske and Houk[57] and then Hopmann[58] mistakenly sketched in a 2D Figure one additional face-to-face interaction with other quinoline moiety (quinoline B) in this favorable transition state structure.

When 3D structure of S1 is qualitatively analyzed, however, only one face-to-face interaction with only one quinoline ring (which is conjugated with the second, quinoline A) can be found. Although the C–H proton is indeed oriented toward pyridazine's N=N bond, the bond length of this C–H proton is not elongated (~1.10 Å), whereas the distances between C atom and H atom with the center of N=N bond is 3.85 and 2.76 Å, respectively, rather suggesting that this interaction is not significantly important.

Note that based on earlier experimental studies from Sharpless group, it was concluded that some π–π stacking may bring important consequences for the reaction enantioselectivity when unsaturated substrates are used.[56] For example, whereas dihydroxylation of styrene afforded the product with 97% ee and vinylcyclohexane afforded the product with 84% ee.[56] The value 84% for nonsaturated substrate can be explained in terms of weak C(sp^3)–H...π hydrogen-bonding interactions between a substrate and a catalyst. Similar C(sp^3)–H...π hydrogen-bonding interactions between the

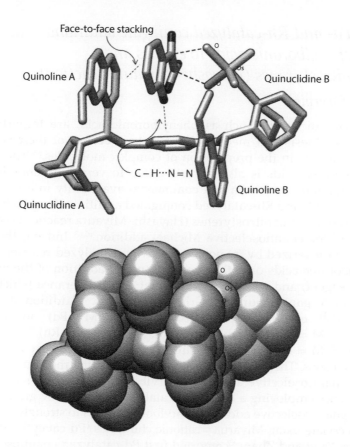

Face-to-face stacking

Quinoline A

Quinuclidine B

C – H···N = N

Quinoline B

Quinuclidine A

Figure 2.10 The most favorable transition state **S1** leading to major (*R*)-1-phenyl-1,2-ethandiol. Noncritical hydrogen atoms are omitted for clarity.

aliphatic chain of the olefin and the quinoline A as well as PYDZ moiety of (DHQD)₂PYDZ were found in dihydroxylation aliphatic *n*-alkenes.[59] The face-to-edge interactions with PYDZ moiety in addition to hydrophobic interactions[60–63] with quinuclidine B were suggested[59] to be responsible for the experimentally observed[31] increase in enantioselectivity with increase of *n*-number from propene up to 1-decene. Because these interactions are relatively very weak in nature, the enantioselectivity value increased from 35% to 85% ee only, corresponding to a ceiling value.

For the so-called "Sharpless mnemonic"[48,49] and "electronic polarizability-based stereochemical"[64,65] models, both qualitative 2D prediction tools for the absolute configuration in this reaction, an interested reader is addressed to the corresponding original publications.

2.4   Pd- and Rh-catalyzed conjugate additions of arylboronic acids to enones and nitrostyrenes (Hayashi–Miyaura reaction)

2.4.1   Overview

Organoboronic acids, such as phenylboronic acid, are important in organic synthesis and medicinal chemistry because of their versatility as synthons in the preparation of complex molecules.[66] The use of organoboronic acids is also attractive due to very high stabilities in protic and aqueous media and commercial availability in a variety of forms. The Pd- and Rh-catalyzed conjugated addition of organoboronic acids to enones and nitrostyrenes (Hayashi–Miyaura reaction) is a powerful tool for enantioselective Michael addition.[67,68] Indeed, the reaction is characterized by (a) no competitive uncatalyzed reaction of the organoboronic acids onto the enone; (b) no 1,2-addition of the organoboron reagent; and (c) a large functional group tolerance which is in contrast to organolithium and Grignard reagents.[69] Addition of arylboronic acids to enones was pioneered by Uemura (M = Pd)[70] and Miyaura (M = Rh,[71] M = Pd[72,73]) in the mid-1990s. Hayashi (M = Rh),[74-77] Miyaura (M = Rh,[77] M = Pd[78]), and Minnaard (M = Pd)[79] reported the first asymmetric versions. The Hayashi group developed the first rhodium-catalyzed highly enantioselective addition of arylboronic acids to α-substituted nitroalkenes employing a Rh/Binap catalyst.[80] The Gutnov group reported the enantioselective addition of arylboronic acids to strongly activated 2-nitroacrylate using Miyarua's cationic chiraphos/Pd catalytic system,[81] whereas Yang and Zhang[82] reported first Pd-catalyzed enantioselective Michael addition of arylboronic acids to nitroalkenes with a broad substrate scope.

In this chapter, origin of enantioselectivity in the conjugate addition of arylboronic acids to β-substituted cyclic enones, nitrostyrenes, and acyclic enones catalyzed by palladium and rhodium complexes is discussed based on computational works of Houk and Stoltz,[83] Yang and Zhang,[82] and Konuklar.[84] The corresponding catalytic reactions (90%–96% ee's) were described in the groups of Stoltz,[83,85-87] Yang and Zhang,[82] and Konno,[88,89] respectively (Scheme 2.10, Chart 2.1).

2.4.2   Catalytic cycle

Miyaura proposed the prototype of the catalytic cycle (M = Pd, Scheme 2.11) in 2004.[73] For the conjugate addition to 3-methyl-2-cyclohexenone outlined in Scheme 2.10, the first step involves transmetalation of cationic Pd(II)–phenylborate complex **A** to generate a phenyl-palladium complex **B**.[83] Complex **B** undergoes ligand exchange to form a more stable

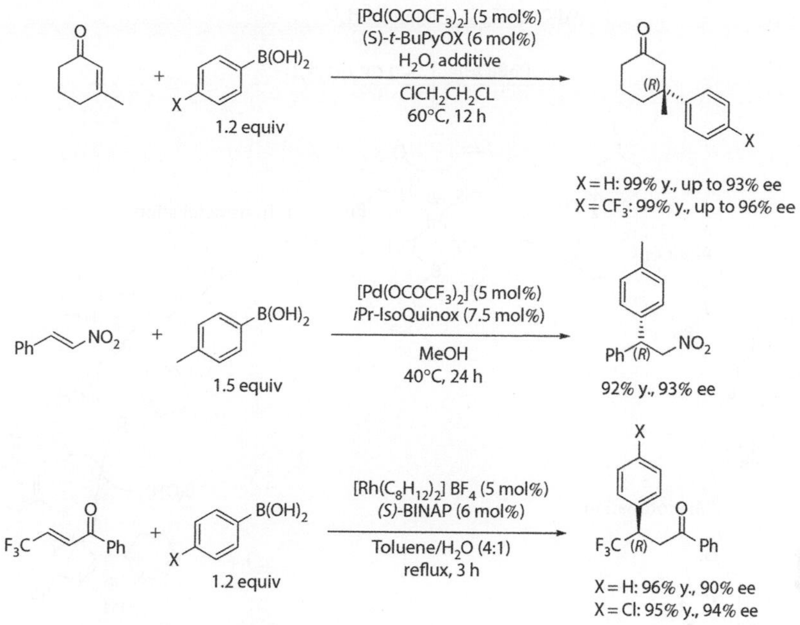

Scheme 2.10 Catalytic reactions.

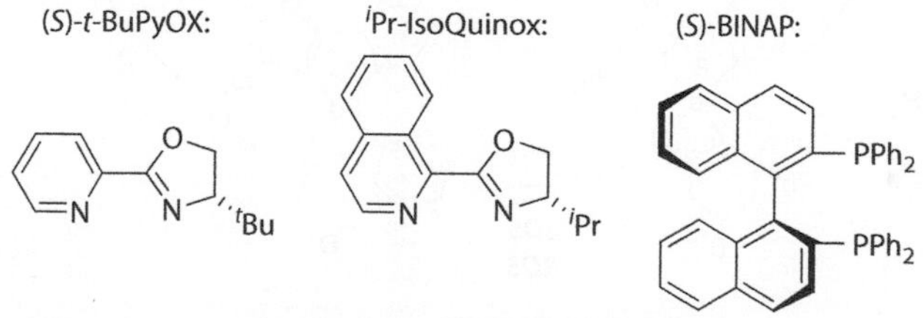

Chart 2.1 Chiral ligands used in the Hayashi–Miyaura reaction.

phenylpalladium–enone complex **C**, in which the palladium binds to the enone oxygen atom. Complex **C** isomerizes to a less stable π complex **D** and then undergoes carbopalladation of the enone to form the new carbon–carbon bond. The carbopalladation step requires an activation free energy of 21.3 kcal·mol⁻¹ (relative to **C**), and is the enantio- and rate-determining transition state. Coordination of one water molecule to **E** leads to an aqua–palladium enolate complex **F**, and finally facile hydrolysis of **F** to afford product complex **G**. Liberation of the product from **G** and coordination with another molecule of phenylboronic acid regenerates complex **A** to complete the catalytic cycle.[83] The computed catalytic cycle demonstrates some similarities with the Pd/bipyridine system in methanol[90,91] and is likely the same as for the conjugate addition to nitrostyrene.

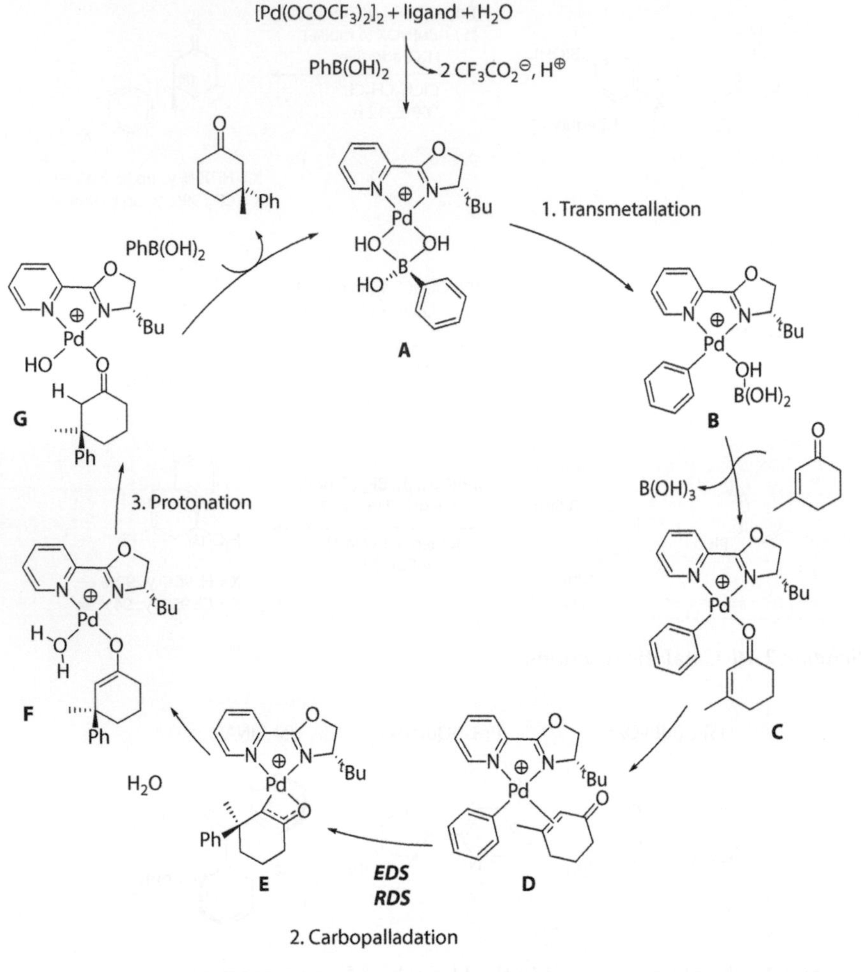

Scheme 2.11 Catalytic cycle for Pd-catalyzed conjugate addition of phenylboronic acid to 3-methyl-2-cyclohexenone.

2.4.3 Catalytic cycle (M = Rh, Scheme 2.12)

Hayashi proposed the prototype of the catalytic cycle for cyclic enone in 2002 based in three identifiable intermediates.[75] Based on this proposition, Konno suggested a plausible cycle for the process shown in Scheme 2.10 underlying that it is only *E*-isomer (more stable) of the substrate that participates in the reaction[89] (Scheme 2.12).

The plausible catalytic cycle for Rh proposed by Konno and further considered in the computations of Konuklar[84] is very similar to Pd, but many details are omitted with respect to a much more detailed and informative cycle of Houk.[83]

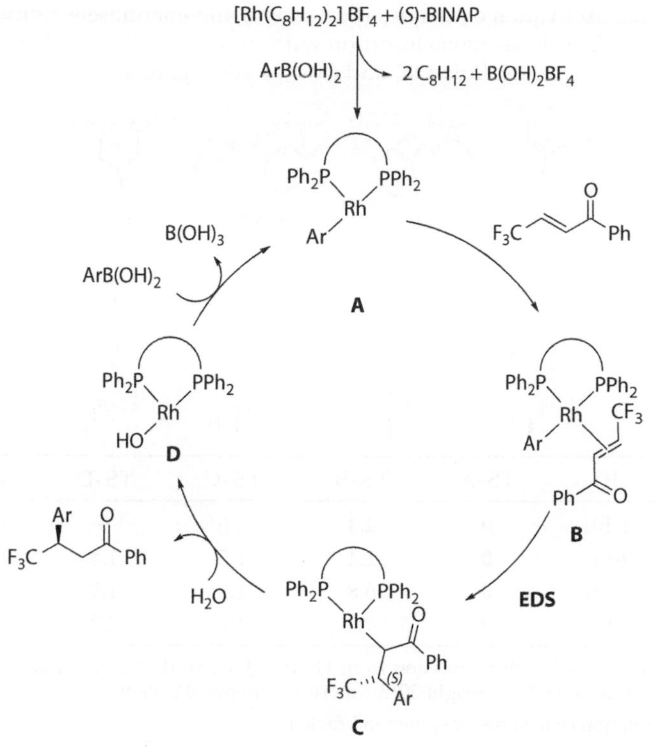

Scheme 2.12 Catalytic cycle for Rh-catalyzed conjugate addition of arylboronic acid to the acyclic enone.

2.4.4   Origin of enantioselectivity (M = Pd)

The enantioselectivity-determining alkene insertion step involves a four-membered cyclic transition state, which adopts a square-planar geometry. When a chiral bidentate ligand, such as PyOx, is employed, there are four possible isomeric alkene insertion transition states (Table 2.2).

Transition state structures **TS-C** and **TS-D** in which phenyl group is located *cis* to the oxazoline, are likely destabilized by steric effects between the R^1 on the ligand and the phenyl group. The best destabilization is achieved when $R^1 = t$-Bu. In contrast, **TS-A** and **TS-B** in which phenyl group is located *trans* to the oxazoline are stabilized by C–H... π interaction between the pyridine hydrogen and phenyl group. The enantioselectivity originates from the energy difference between transition states **TS-A** and **TS-B**. For $R^1 = t$-Bu, **1-TS-B** is likely destabilized by steric effects originating from the t-Bu group. In contrast, no ligand–substrate steric repulsions are observed in **1-TS-A**, in which the cyclohexenone is *anti* to the t-Bu. **1-TS-A** is also stabilized by a weak hydrogen bond between the

Table 2.2 Activation enthalpies (kcal·mol⁻¹) and enantioselectivities of 3-methyl-2-cyclohexenone insertion with (S)-*t*-BuPyOX, (S)-*i*-PrPyOX, (S)-*i*-BuPyOX, and (S)-PhPyOx ligands

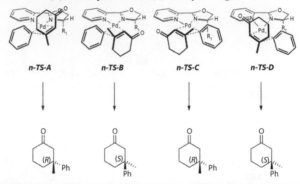

TS	R¹	TS-A	TS-B	TS-C	TS-D	ee (%)[a]
1	*t*-Bu	0	**2.3**	2.6	3	94 (93)
2	*i*-Pr	0	**1.1**	1.5	1.4	67 (40)
3	*i*-Bu	0	**0.8**	1.9	1.9	52 (24)
4	Ph	0	**1**	1.6	2.2	65 (52)

Source: Reprinted with permission from Holder, J. C. et al., *J. Am. Chem. Soc.,* 135, 14996–15007. Copyright 2013 American Chemical Society.

[a] Experimental ee's values are given in brackets.

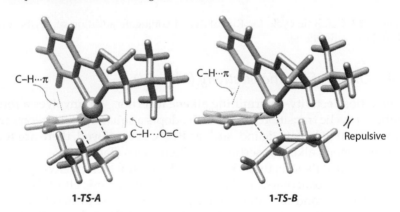

Figure 2.11 The most favorable transition states leading to major (R)- and minor (S)-product. Optimized at BP86/SDD(Pd)/6-31G*(all others) level of theory.

carbonyl oxygen and the hydrogen geminal to the *t*-Bu group on the oxazoline. The 3D structures of **TS-A** and **TS-B** are shown in Figure 2.11.

The chiral induction for the nitrostyrene Michael addition can be explained using the similar model (Figure 2.12).[82] The nucleophilic aryl group coordinated to the Pd nitrostyrene is *cis* to the oxazoline. The

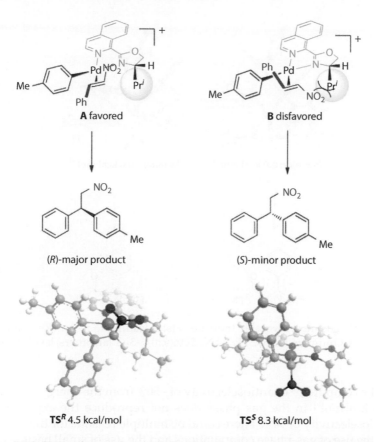

Figure 2.12 Stereochemical model for Pd-catalyzed enantioselective Michael addition of arylboronic acid to nitroalkene. (Adapted with permission from He, Q. et al., *Org. Lett.*, 17, 2250–2253. Copyright 2015 American Chemical Society.)

sterically disfavored intermediate (**B**), in which the substrate coordinates to the palladium with the $-NO_2$ group oriented downward (*si*-face) close to the isopropanyl group of IsoQuinox, leads to the optically minor product (*S*)-3. DFT calculations showed that the energy of the transition state **Ts^S** was 8.3 kcal·mol^{-1}. The intermediate **A** is more sterically favored. The energy of the corresponding transition state **Ts^R** is 4.5 kcal·mol^{-1}.

2.4.5   *Origin of enantioselectivity (M = Rh)*

Similarly to Pd, the enantioselectivity-determining alkene insertion step involves a four-membered cyclic transition state, which adopts a distored square-planar geometry. For BINAP, there are two possible isomeric alkene insertion transition states, leading to *R* and *S*-products, respectively, as shown in Figure 2.13.

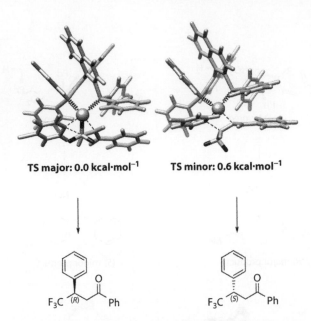

TS major: 0.0 kcal·mol⁻¹   TS minor: 0.6 kcal·mol⁻¹

Figure 2.13 The most favorable transition states leading to major (*R*)- and minor (*S*)-product. Optimized at B3LYP/LANL2DZ(Rh)/6-31G*(all others) level of theory.

The computed[84] enantioselectivity of ~50% from the energy difference of 0.6 kcal·mol⁻¹ in the gas phase does not reproduce the experimental enantioselectivity of 94%. There could be multiple reasons for this starting from the use of gas-phase computations and the use of small basis set and old functional and finishing by wrong reaction mechanism.

2.5   Rh-catalyzed asymmetric hydroboration of vinylarenes

2.5.1   Overview

Olefin hydroboration, which is the addition of a B–H bond across C=C bond, was first discovered by H. C. Brown in 1956[92] and Köster in 1958.[93] Typically, the reaction does not require a catalyst and the simple borane reagent (e.g., B_2H_6, BH_3·THF, BH_3·SMe$_2$, BH$_2$Cl·Et$_2$O, thexylborane, disiamylborane, and 9-BBN) or boranes bearing electron-withdrawing substituents (e.g., Piers's borane $B(C_6F_5)_2$[94]) react rapidly even at room temperature to afford, after oxidation, the linear anti-Markovnikov products. The reaction can be remarkably C=C/C=O chemoselective for terminal alkenes containing carbonyl groups.[95] For hydroborating reagents in which boron atom is bonded to heteroatoms such as oxygen or nitrogen (e.g., catechol-borane, pinacolborane, and ephidrineborane) elevated temperatures are needed for reaction to occur.[96–98] Although the uncatalyzed hydroboration–oxidation of alkenes usually affords the anti-Markovnikov products, the catalyzed versions can be induced to produce either Markovnikov or anti-Markovnikov products.[99] On the contrary to metal-free, the metal-catalyzed anti-Markovnikov reaction with diorganyloxyboranes proceeds to completion within minutes at room temperature.[100] The regioselectiv-ity and enantioselectivity obtained with a catalyst depend on the ligands attached to the metal, form of the metal precursor used (i.e., neutral or cat-ionic complex), nature of the metal used (e.g., Rh or Ir), steric and electronic properties of the reacting alkene, nature of the borane used (e.g., catechol-borane vs. pinacolborane), additives, and other parameters. For example, catalytic hydroboration of perfluoroalkylalkenes with cationic[101,102] and neutral[101] rhodium complexes allows for selective access to Markovnikov and anti-Markovnikov alcohol products, respectively, after hydrobora-tion with catechol- and pinacolboranes, followed by oxidation with alka-line hydrogen peroxide. Catecholborane usually affords Markovnikov product, whereas pinacolborane affords a mixture of Markovnikov and anti-Markovnikov products in the cationic Rh-BINAP-catalyzed hydroboration–oxidation of vinylarenes.[103] In contrast, the selectivities are similar for these reagents, when cationic Rh-QUINAP complexes are used (still, however, catecholborane gives better regio- and enantioselectivity).[103] Substitution of Rh by Ir in the hydroboration of vinylarenes[104] or meso substrates[105] leads to complete reverse of the sense of enantioinduction and/or regioselectivity.

The first transition-metal-promoted (M = Ir, Co) reactions of simple boron hydrides to alkynes was reported by Sneddon in the beginning of 1980s.[106,107] The first olefin hydroboration (M = Rh) with diorganyloxy-boranes was developed in 1985 by Männig and Nöth.[108] The first enan-tioselective (Rh-catalyzed) olefin hydroboration with these and similar boranes were developed by Suzuki,[109] Evans,[110] Burges,[111] Hayashi and Ito,[112] and Lloyd-Jones and J. M. Brown[113] in the end of 1980s. The metal-catalyzed asymmetric hydroboration of olefins with hydroborating agents in which boron atom is bonded to heteroatoms such as oxygen or nitrogen is a straightforward method to obtain chiral boronic esters with up to 98% ee.[114–116] The later can be converted into a variety of functionalities includ-ing alcohols, carboxylic acids, amines, halides, and homologated materi-als[117–120] and with retention of stereochemistry.[115]

In this chapter, origin of enantioselectivity in the hydroboration/oxidation of vinylarenes with catecholborane by cationic rhodium com-plexes is discussed based on computational work of Fernández and Bo as

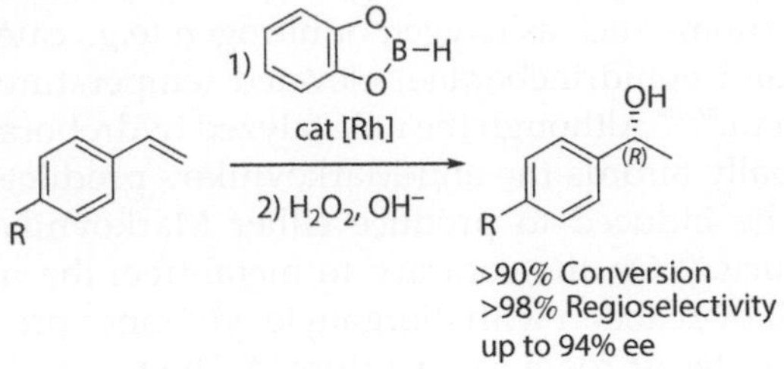

Scheme 2.13 Reactions.

Entry	Catalytic system	R	T(°C)	ee (%)
1	[Rh(COD){(R)-BINAP}]BF$_4$	H	25	57
2	[Rh(COD){(R)-QUINAP}]BF$_4$	H	25	92
3	[Rh(COD){(R)-QUINAP}]BF$_4$	Me	25	89
4	[Rh(COD){(R)-QUINAP}]BF$_4$	OMe	25	94
5	[Rh(COD){(R)-QUINAP}]BF$_4$	F	25	80
6	[Rh(COD){(R)-QUINAP}]BF$_4$	Cl	25	78
7	[Rh(COD){(R)-QUINAP}]BF$_4$	CF$_3$	25	45
8	[Rh(COD)$_2$]BF$_4$/(R)-PYPHOS	H	25	69
9	[Rh(COD)$_2$]BF$_4$/(R)-PYPHOS	H	0	90

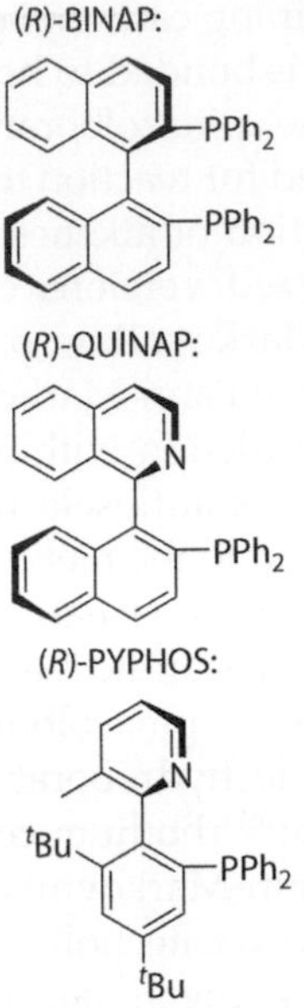

shown in Scheme 2.13.[103,121] The corresponding catalytic reactions leading to branched Markovnikov products (>98% selectivity) were described in the groups of Hayashi and Ito,[122] J. M. Brown,[123] and Chan.[124]

The following conclusions can be drawn from experimental observations (Scheme 2.13): higher enantioselectivities are observed for heterotopic P,N ligands (QUINAP, PYPHOS) than for homotopic P,P ligands (BINAP); the enantioselectivity is temperature-dependent with heterotopic P,N ligands in the asymmetric induction of styrene (Scheme 2.13, entries 8 and 9); and the electronic nature of the 4-substituted styrene influences stereoselectivity to the extent that ee values seem to obey a linear free energy relationship with the Hammett constants (Scheme 2.13, entries 2–7). Note that when pinacolborane was used instead of catecholborane for the same reaction catalyzed by [Rh(COD){(R)-BINAP}]BF$_4$, two products Markovnikov and anti-Markovnikov were observed in 1:1 ratio, and the branched product was obtained with the opposite absolute configuration of 18%.[103] On the other hand, the same sense of enantioselection was observed with the QUINAP ligand.[103]

2.5.2 Catalytic cycle

The presently accepted catalytic cycle follows inner-sphere mechanism (Scheme 2.14).[115] The inner-sphere mechanism within which the

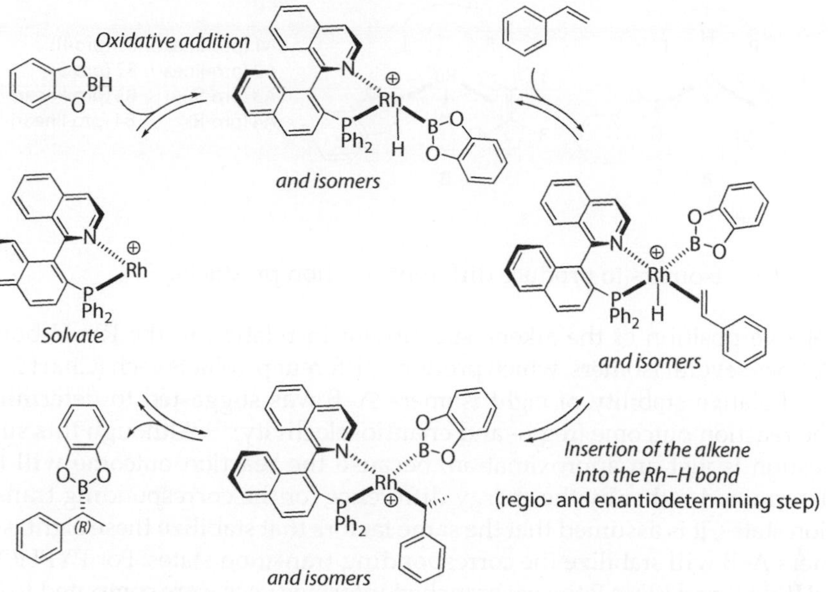

Scheme 2.14 Presently accepted catalytic cycle for the cationic Rh-catalyzed asymmetric hydroboration of vinylbenzene leading to Markovnikov product.

unsaturated substrate is π-coordinated on the metal is particularly supported by two facts: C=C versus C=O chemoselectivity[108] and relatively high catalyst loadings. For example, Noyori-type catalysts that operate in the outer sphere (no substrate coordination on the metal) have the opposite C=O versus C=C chemoselectivity and perform under very low catalyst loadings.[125] An initial oxidative addition of the B–H bond at the cationic metal center is followed by the π-coordination of the unsaturated C–C bond to the metal center. Subsequent migration of hydride would occur to either the terminal or the substituted carbon atom. Alternatively, migratory insertion of the alkene to Rh–B bond may occur (C–B coupling) prior to reductive C–H elimination,[126] but this possibility is not generally accepted,[115] therefore not shown in Scheme 2.14. Finally, reductive C–B elimination from the metal center would yield the hydroborated products and regenerate the metal catalyst.

2.5.3 Enantioselectivity

Fernández and Bo suggested that the key intermediate responsible for both regio- and enantioselectivity is pentacoordinated H–Rh–(ligand)–catecholborane–styrene complex, where styrene is coordinated *trans* to the naphthylpyridine moiety and the hydride is in axial position.[103,119] The

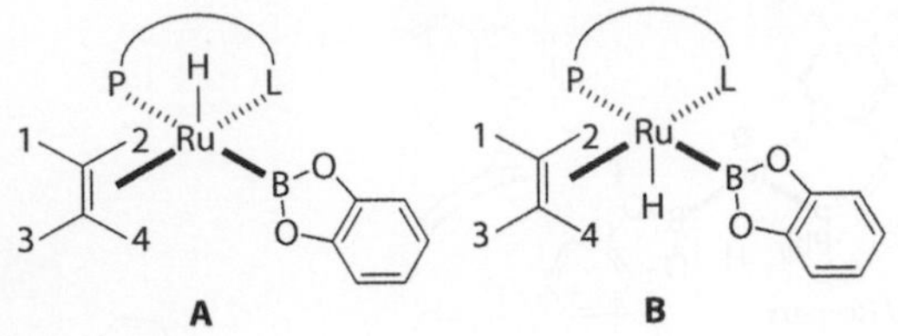

A1 (pro-linear)   B1 (pro-R)
A2 (pro-linear)   B2 (pro-S)
A3 (pro-S)   B3 (pro-linear)
A4 (pro-R)   B4 (pro-linear)

L = N, P

Chart 2.2 Isomers to produce different reaction products.

relative position of the alkene substituent in relation to the Rh–H bond defines several isomers, which produce different products each (Chart 2.2).

Relative stability of eight isomers A–B was suggested to determine the reaction outcome (regio- and enantioselectivity).[121] Although this suggestion is just an approximation, because the reaction outcome will be determined rather by the energy difference for the corresponding transition states, it is assumed that the same factors that stabilize these eight isomers A–B will stabilize the corresponding transition states. For PYPHOS, QUINAP and BINAP the probranched intermediates were computed to be uniformly more stable than prolinear intermediates. For all three ligands, the most stable isomer was computed to be B1, the one that led to the *R*-enantiomer in agreement with experimental observation. For PYPHOS and QUINAP, the second most stable isomer was computed to be pro-*S* B2. The computed QM/MM energy difference of 3.5 kcal·mol^{-1} for PYPHOS and 4.1 kcal·mol^{-1} for QUINAP is in agreement with experimental observation that higher ee's are obtained for QUINAP at 25°C (Scheme 2.13). In general in PYPHOS, because the difference between B1 and B2 is smaller than in QUINAP, a lower temperature is needed to obtain the same ee. For more-crowded BINAP, the second most stable isomer was computed to be pro-*S* A3, 0.3 kcal·mol^{-1} above B1, qualitatively indicating a lower stereo-differentiation, in agreement with experiment.

The difference between B1 and B2 (PYPHOS, QUINAP) arose from the different π–π interactions between the substrates and the catalyst. There are in total three types of π–π interactions in structures B1, B2, and A3: π1 interaction that are always present between a ligand aromatic rings, π2 interaction between the substrate (phenyl aromatic ring) and the ligand present in B1 and π3 interaction between the substrate (phenyl aromatic ring) and catecholborane (five-membered ring) present in B2. The always present π1 interaction does not determine the reaction outcome (Figure 2.14). The slipped-parallel π2 interaction is stabilizing, whereas π3 interaction is destabilizing. Therefore all three ligand, substrate, and catecholborane are responsible for the stereoinduction: Stabilizing π–π interaction in B1 and destabilizing π–π interaction in B2 both play a key role in enatiodifferentiation. Note that destabilizing π–π interaction in B2

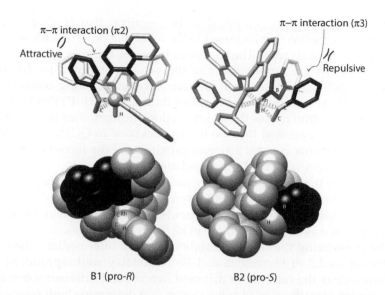

Figure 2.14 The most stable isomers of penta-coordinated H–Rh–(ligand)–catecholborane–styrene complex leading to *R*- and *S*-configured products under the assumption that the process proceed via insertion of the double bond into the Rh–H bond.

is absent if pinacolborane is considered instead of catecholborane, and a new $CH_3 \ldots \pi$ appears between the substrate and pinacolborane, thus explaining smaller ee's with this reagent.[103]

What is the evidence that π2 interaction is stabilizing and π3 interaction is destabilizing? The relative strength of each π–π interaction was evaluated by considering model systems (Chart 2.3).

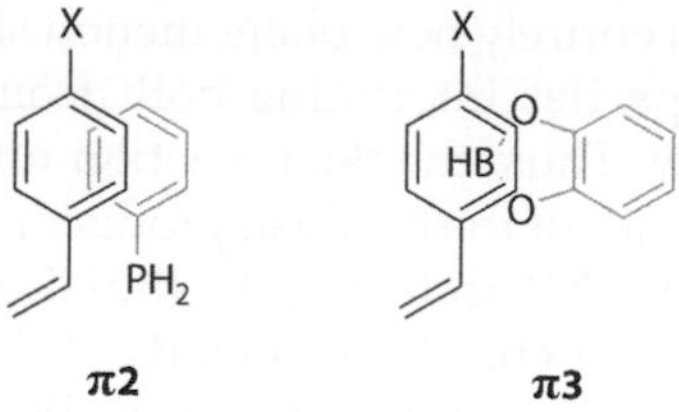

	π2		π3	
X	IE	d	IE	d
OMe	−3.9	3.314	0.2	4.641
H	−4.6	3.539	−0.2	4.525
Cl	−5.1	3.375	−0.7	4.661

Chart 2.3 Interaction energy (IE in kcal•mol^{-1}), distance between ring centroids (*d* in Å) in the model systems.

For π2, the IE was computed to be negative, that is, there was a stabilizing interaction that increased in the following order: *p*-OMe-styrene < styrene < *p*-Cl-styrene. The distances between the ring centroids (3.4 Å on average) were quite similar to the values computed for B1 complexes in QUINAP. The shorter distance between the ring centroids in the Rh-QUINAP complexes (3.5 Å) than those in the Rh-PYPHOS (4.1 Å) complexes therefore makes the π2 interaction stronger, thus adding stabilization. It was suggested that the ligand backbone in QUINAP provides a conformation that enables a greater proximity of the ligand to the substrate, which stabilizes the B1 intermediate more.[121]

For π3, the IE was computed to be positive or only slightly negative, but the relative trend was similar *p*-OMe-styrene < styrene < *p*-Cl-styrene. The equilibrium distance between ring centroids in the model systems for π3 (4.6 Å on average) was longer than those in π2 (3.4 Å on average).

By considering the relative stability of a key intermediate, the computational model of Fernández and Bo reasonably well reproduced the performance of the catalyst for different ligands and different substrates, and allowed identification of some factors that determine both regio- and enantiodifferentiation.

However, one must keep in mind that not only TS's were not computed in this study, but also the number of analyzed catalytic pathways has been artificially and significantly reduced.

2.6 Mechanism of autoamplifying Soai reaction

2.6.1 Introduction

In 1995 Soai et al. discovered a reaction that remains so far essentially the only straightforward example of a chemical process that leads to amplification of the catalyst chirality (Scheme 2.15).[127] Intensive research of the properties of this entirely new phenomenon showed that it features numerous unique properties interesting both from the theoretical and practical points of view. Thus, the Soai reaction offers the possibility of bringing the optical purity of the catalyst/product to almost absolute perfection by repeated runs,[128–131] inducing the initial chirality of the sample by microscopic amounts of chiral compounds,[128–131] and even of generating scalemic samples from nonchiral precursors.[132–139]

Physicochemical understanding of the amplification and autoamplification phenomena is based on the kinetic Scheme formulated by Frank in 1953 (Scheme 2.16).[140] Formal kinetic treatment of the equations shown in Scheme 2.16 allows us to conclude that if the heterochiral dimeric product is removed from the flux of catalysis (e.g., Equation 5 in Scheme 2.16), in a flow reactor the autocatalysis (Equations 3 and 4 in Scheme 2.16) will

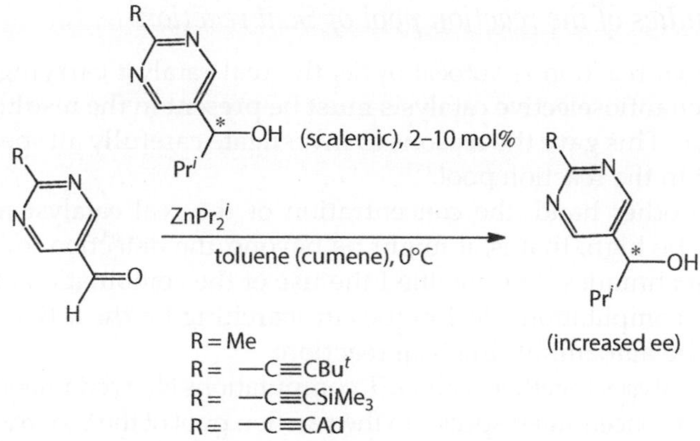

Scheme 2.15 Soai autoamplification reaction.

$$(1)\ A \xrightarrow{k_0} R \quad (2)\ A \xrightarrow{k_0} S \quad (3)\ R + A \xrightarrow{k_1} 2R$$

$$(4)\ S + A \xrightarrow{k_1} 2S \quad (5)\ R + S \xrightarrow{k_2} RS$$

Scheme 2.16 Frank's Scheme for the asymmetric amplification where A is substrate; R, S are enantiomer products; and RS is heterochiral dimer.

develop any initial ee of the catalyst/product (no matter how small it may be) to a reaction pool consisting of enantiomerically pure catalyst/product of the same handedness even in the presence of a slower background reaction (Equations 1 and 2 in Scheme 2.16).[141–154] The smaller the initial enantiomeric excess, the longer the time required for the development of asymmetry. In a closed reactor, the same process can be obscured by the fact that the substrate may be exhausted earlier than the characteristic time of asymmetry development is achieved.[154]

All further extensive work on the physicochemical background of the asymmetry amplification is based on various modifications of the initial Frank's Scheme.[141–154] Hence, there is no lack of understanding of this phenomenon from the kinetic side of the problem.

However, having in mind the evident virtues of conscious creation of various self-amplifying reactions, much more important is the problem of the essential structural requirements that would allow the actual realization of a self-amplifying scenario. From this point of view the studies of the mechanism of autoamplification in the Soai reaction are timely and highly interesting.

2.6.2   Studies of the reaction pool of Soai reaction

Since the Soai reaction is autocatalytic, the real catalyst carrying out the flux of the enantioselective catalysis must be present in the resulting reaction mixture. This gave the reasons to investigate carefully all species that are present in the reaction pool.[155]

On the other hand, the concentration of the real catalyst must not necessarily be high, that is, it might be beyond the detection level of the analytical techniques. This justified the use of the combination of experimental and computational techniques in searching for the active catalytic species in the autoamplifying Soai reaction.

NMR analyses together with DFT computations showed unequivocally that the most concentrated species in the reaction pool of the Soai reaction are homo- and hetero-square dimers **2** existing in a dynamic equilibrium with monomers **1**.[156–160] Homo- and hetero-square dimers **2** are present in the reaction pool in practically equal concentrations that excludes the possibility of involvement of a reservoir mechanism in the catalytic autoamplification.[161]

Experimentally determined activation barrier of the dissociation (19.1 kcal mol^{-1})[156] was used to choose the computational technique for an adequate description of the Soai reaction. As can be seen by analysis of the computed thermodynamic parameters shown in Scheme 2.17, whereas the B3LYP functional provides a reasonable approximation of the dissociation barrier, the use of M05-2X with the same basis strongly overestimates the strength of the dimers.

Although the square dimers **2** are undoubtedly the most concentrated species in the reaction pool of the Soai reaction, they are unlikely candidates for an autoamplifying catalyst, since the strict substrate specificity of the reaction implies participation of the nitrogen atoms of pyrimidinyl rings in the structure of the active catalyst.

Hence, it has been suggested that macrocyclic dimers **3** could be considered as the probable catalytic species, however they are thermodynamically disfavored compared to the square dimers (Scheme 2.18).

Nevertheless, the formation of the macrocyclic units is considered to be essential so far in both suggested catalytic cycles. Thus, Ercolani and Shaffiano accepted the M05-2X computational results implying that coordination of two ZnPr$^i_2$ molecules to the oxygen atoms of the macrocycle makes it relatively stable (notice, however, the tremendous difference between the relative stabilities obtained by different functionals). In the later work, it has been shown that the macrocyclic units are likely to be reversibly generated by aggregation of two square dimers (Scheme 2.19).[154,155]

The B3LYP results predict endogonic formation of tetramers **4** at ambient temperature that becomes slightly exogonic at 193 K. This corresponds well to the experimental observations. On the other hand, the

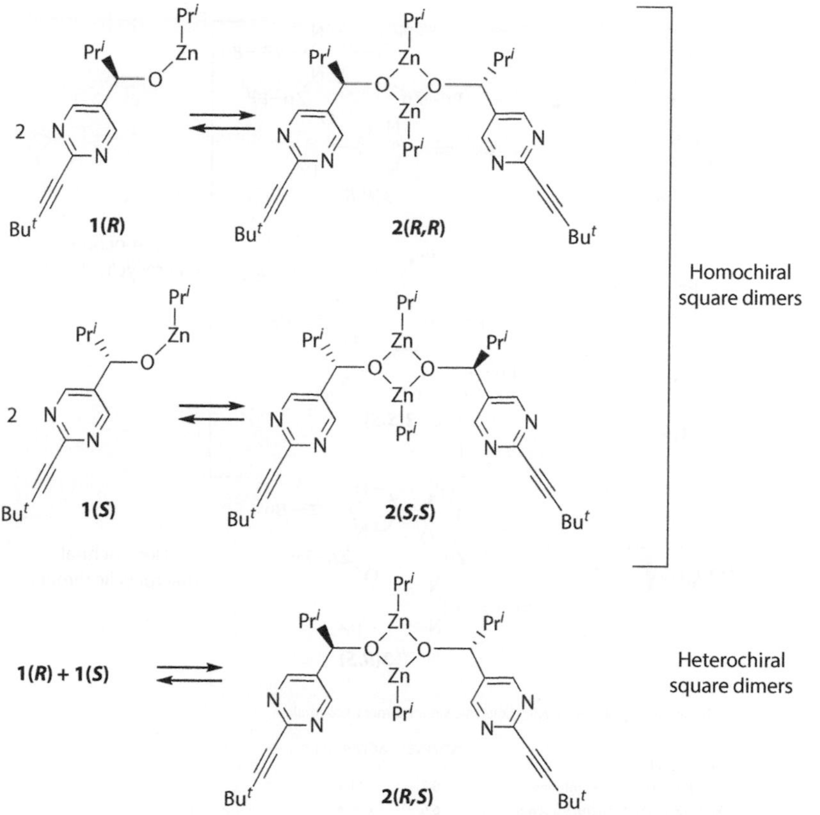

Homochiral square dimers

Heterochiral square dimers

Scheme 2.17 Most concentrated species in the reaction pool of the Soai reaction are homo- and heterochiral square dimers 2. (Reproduced with permission from Gridnev, I. D. et al., *ACS Catal.*, 2, 2137–2149, 2015.)

Free energy of dimerization, kcal/mol	ΔE(ZPVE)	ΔG(298.15)	ΔG(273.15)	ΔG(193.15)
B3LYP/6-31G*(cpcm, toluene)	−30.6	−16.0	−17.2	−21.1
M05-2X/6-31G*(cpcm, toluene)	−54.0	−29.2	−30.9	−38.1

calculations by M05-2X functional evidently strongly overestimate the strengths of the Zn–O and Zn–N bond that would result experimentally in the stupendous stability of the tetrameric species that is not observed in reality.[155]

The evident and significant advantage of B3LYP in describing the reaction pool of the Soai reaction made us to skip the description of the mechanism suggested by Ercolani and Shaffiano which is essentially based on the M05-2X calculations in gas phase without accounting for the entropy.[162–165] Besides, its careful analysis involves competition of

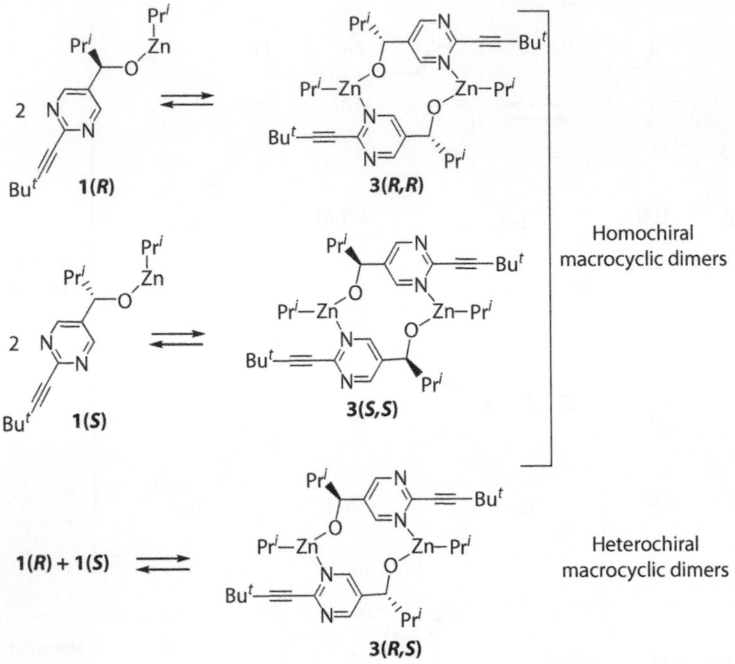

Scheme 2.18 Formation of macrocyclic dimers **3** and energies of the most stable conformers relative to the square dimers. (Reproduced with permission from Gridnev, I. D. et al., *ACS Catal.*, 2, 2137–2149, 2015.)

numerous catalytic cycles with sophisticated geometries of the crucial intermediates and transition states, as well as far from evident stereo-chemical preferences. Hence, an interested reader is addressed to the corresponding original publications cited here.[162–165]

2.6.3　Structure of the catalyst and computations of the catalytic cycle of Soai reaction

Detailed conformational analysis of the macrocyclic tetrameric species formed by dimerization of the Zn–O–Zn–O square dimers and of their ZnPr$_2^i$ adducts revealed the structural divergence of the homo- and

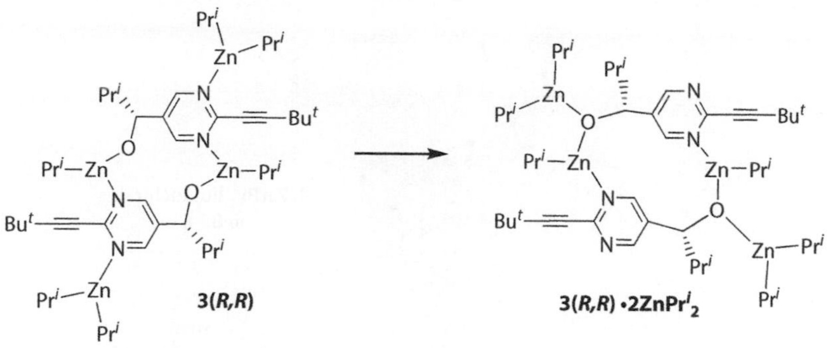

Relative stabilities of $ZnPri_2$ complexes, kcal/mol	ΔE(ZPVE)	ΔG(298.15)
B3LYP/6-31G* (cpcm, toluene)	14.5	21.9
M05-2X/6-31G* (cpcm, toluene)	−1.4	3.8

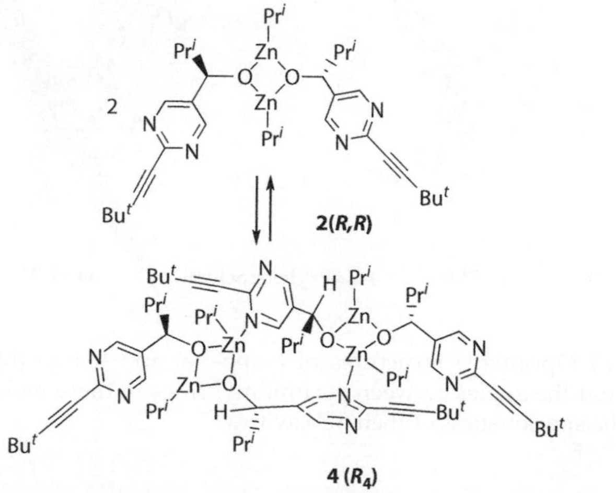

Energy of tetramerization, kcal/mol	ΔE(ZPVE)	ΔG(298.15)	ΔG(273.15)	ΔG(193.15)
B3LYP/6-31G* (cpcm, toluene)	−30.6	5.8	4.1	−1.5
M05-2X/6-31G* (cpcm, toluene)	−42.2	−24.0	−25.4	−29.3

Scheme 2.19 Relative stabilities of $ZnPr^i_2$ complexes and thermodynamic parameters of tetramerization computed with two different functionals. (Reproduced with permission from Gridnev, I. D. et al., *ACS Catal.*, 2, 2137–2149, 2015.)

heterochiral species. Homochiral tetramers are exclusively formed in a specific *brandyglass* conformation (not available on the dimer level) with almost orthogonal pyrimidinyl rings forming a 3D cavity that is virtually unaffected by the formation of a $ZnPr_2{}^i$ adduct (Figure 2.15, top). Similar conformation of the heterochiral macrocyclic tetramer is significantly less spacious and is practically closed upon coordination of a molecule of $ZnPr_2{}^i$ (Figure 2.15, bottom).[155]

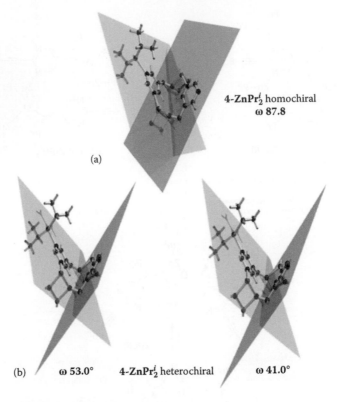

Figure 2.15 Optimized structures of homo- (a) and hetero (b) chiral adducts 4•ZnPr$_2^i$ and the angles between pyrimidinyl rings in these molecules that characterize the spaciousness of their 3D cavities.

The 3D cavity observed in the ZnPr$_2^i$ adduct of the homochiral *brandyglass* tetramer forms an ideal chiral pocket for coordination of the aldehyde followed by perfectly enantioselective alkylation yielding monomeric alcoholate of the same handedness as the tetrameric catalyst (Figure 2.16).[155] A possibility of the opposite way of the aldehyde coordination is completely excluded, because switching of the prochiral planes of the coordinated aldehyde would replace the oxygen atom in the proximity of Zn with hydrogen, and activation would become impossible.

A significantly less spacious cavity in the heterochiral *brandyglass* tetramer practically disappears upon the coordination of ZnPr$_2^i$ which excludes the heterochiral species from the flow of catalysis and leads to the realization of the Frank's Scheme for the amplification of chirality.[154]

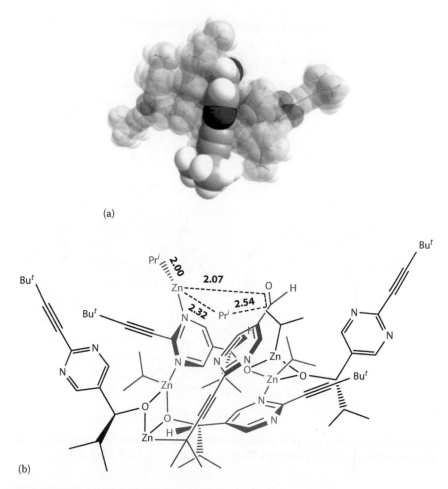

Figure 2.16 (a,b) Transition state for the direct transfer of the isopropyl group from the N-bound $ZnPr2_i$ computed for the alkylation of aldehyde with the catalyst **4** (B3LYP/6-31G [CPCM, toluene]). The importance of the specifically shaped substrate accurately fitting to the cavity is illustrated on the space-filling model. (Reproduced with permission from Gridnev, I. D. et al., *ACS Catal.*, 2, 2137–2149, 2015.)

The catalytic cycle (Scheme 2.20) involves reversible association of the square dimers **2** into the tetramer **4**, N-coordination of diisopropyl zinc, capture of the aldehyde molecule in the 3D cavity, alkylation, and dissociation of the monomeric alcoholate that amplifies the chiral pool of the reaction.[155]

It has been shown recently that the observation of a transient intermediate with the acetal structure in the reaction pool of Soai reaction[166]

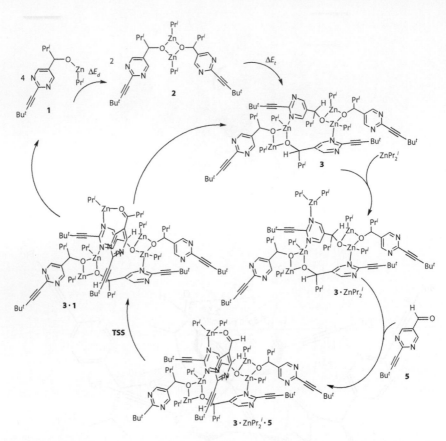

Scheme 2.20 Catalytic cycle of Soai Reaction involving tetrameric catalyst. (Reproduced with permission from Gridnev, I. D. et al., *ACS Catal.*, 2, 2137–2149, 2015.)

can be explained accepting the catalytic cycle shown in Scheme 2.20, since the formation of a monomeric product is an essential requirement for the kinetic stability of the acetal species.[167]

References

1. Noyori, R.; Hashiguchi, S. Asymmetric Transfer Hydrogenation Catalyzed by Chiral Ruthenium Complexes. *Acc. Chem. Res.* 1997, *30*, 97–102.
2. Ikariya, T.; Blacker, A. J. Asymmetric Transfer Hydrogenation of Ketones with Bifunctional Transition Metal-Based Molecular Catalysts. *Acc. Chem. Res.* 2007, *40*, 1300–1308.
3. Ikariya, T.; Murata, K.; Noyori, R. Bifunctional Transition Metal-based Molecular Catalysts for Asymmetric Syntheses. *Org. Biomol. Chem.* 2006, *4*, 393–406.

4. Ikariya, T.; Gridnev, I. D. Bifunctional Transition Metal-based Molecular Catalysts for Asymmetric C–C and C–N Bond Formation. *Chem. Rec.* 2009, *9*, 106–123.
5. Ikariya, T.; Gridnev, I. D. Bifunctional Transition Metal-based Molecular Catalysts for Asymmetric C–C and C–N Bond Formation. *Topics in Cat.* 2010, *53*, 894–901.
6. Hashiguchi, S.; Fujii, A.; Takehara, J.; Ikariya, T.; Noyori, R. Asymmetric Transfer Hydrogenation of Aromatic Ketones Catalyzed by Chiral Ruthenium (II) Complexes. *J. Am. Chem. Soc.* 1995, *117*, 7562–7563.
7. Takehara, J.; Hashiguchi, S.; Fujii, A.; Inoue, S.; Ikariya, T.; Noyori, R. Amino Alcohol Effects on the Ruthenium(II)-Catalysed Asymmetric Transfer Hydrogenation of Ketones in Propan-2-ol. *Chem. Commun.* 1996, 233.
8. Gao, J.-X.; Ikariya, T.; Noyori, R. A. Ruthenium(II) Complex with a C_2-Symmetric Diphosphine/Diamine Tetradentate Ligand for Asymmetric Transfer Hydrogenation of Aromatic Ketones. *Organometallics* 1996, *15*, 1087–1089.
9. Fujii, A.; Hashiguchi, S.; Uematsu, N.; Ikariya, T.; Noyori, R. Ruthenium(II)-Catalyzed Asymmetric Transfer Hydrogenation of Ketones Using a Formic Acid-Triethylamine Mixture. *J. Am. Chem. Soc.* 1996, *118*, 2521–2522.
10. Uematsu, N.; Fujii, A.; Hashiguchi, S.; Ikariya, T.; Noyori, R. Asymmetric Transfer Hydrogenation of Imines. *J. Am. Chem. Soc.* 1996, *118*, 4916–4917.
11. Hashiguchi, S.; Fujii, A.; Haack, K.-J.; Matsumura, T.; Ikariya, T.; Noyori, R. Kinetic Resolution of Racemic Secondary Alcohols by Ru^{II}-Catalyzed Hydrogen Transfer. *Angew. Chem. Int. Ed. Engl.* 1997, *36*, 288–290.
12. Matsumura, K.; Hashiguchi, S.; Ikariya, T.; Noyori, R. Asymmetric Transfer Hydrogenation of α,β-Acetylenic Ketones. *J. Am. Chem. Soc.* 1997, *119*, 8738–8739.
13. Ikariya, T.; Hashiguchi, S.; Murata, K.; Noyori, R. Preparation of Optically Active (*r,r*)-Hydrobenzoin from Benzoin or Benzyl. *Org. Synth.* 2005, *82*, 10–17.
14. Gridnev, I. D.; Watanabe, M.; Wang, H.; Ikariya, T. Mechanism of Enantioselective C–C Bond Formation with Bifunctional Chiral Ru Catalysts: NMR and DFT Study. *J. Am. Chem. Soc.* 2010, *132*, 16637–16650.
15. Hasegawa, Y.; Watanabe, M.; Gridnev, I. D.; Ikariya, T. Enantioselective Direct Amination of α-Cyanoacetates Catalyzed by Bifunctional Chiral Ru and Ir Amido Complexes. *J. Am. Chem. Soc.* 2008, *130*, 2158–2159.
16. Hasegawa, Y.; Gridnev, I. D.; Ikariya, T. Enantioselective and Z/E-Selective Conjugate Addition of α-Substituted Cyanoacetates to Acetylenic Esters Catalyzed by Bifunctional Ruthenium and Iridium Complexes. *Angew. Chem. Int. Ed.* 2010, *49*, 8157–8160.
17. Hasegawa, Y.; Gridnev, I. D.; Ikariya, T. Mechanistic Consideration of Asymmetric C–N and C–C Bond Formations with Bifunctional Chiral Ir and Ru Catalysts. *Bull. Chem. Soc. Jpn.* 2012, *85*, 316–334.
18. Ishida, N.; Sawano, S.; Murakami, M. Synthesis of 3,3-Disubstituted [Small Alpha]-Tetralones by Rhodium-Catalysed Reaction of 1-(2-haloaryl) Cyclobutanols. *Chem. Commun.* 2012, *48*, 1973–1975.
19. Seiser, T.; Cramer, N. Rhodium(I)-Catalyzed 1,4-Silicon Shift of Unactivated Silanes from Aryl to Alkyl: Enantioselective Synthesis of Indanol Derivatives. *Angew. Chem., Int. Ed.* 2010, *49*, 10163–10167.

20. Shigeno, M.; Yamamoto, T.; Murakami, M. Stereoselective Restructuring of 3-Arylcyclobutanols into 1-Indanols by Sequential Breaking and Formation of Carbon–Carbon Bonds. *Chem.* 2009, *15*, 12929–12931.
21. Murakami, M.; Tsuruta, T.; Ito, Y. Lactone Formation by Rhodium-Catalyzed C–C Bond Cleavage of Cyclobutanone. *Angew. Chem., Int. Ed.* 2000, *39*, 2484–2486.
22. Murakami, M.; Itahashi, T.; Amii, H.; Takahashi, K.; Ito, Y. New Domino Sequences Involving Successive Cleavage of Carbon–Carbon and Carbon–Oxygen Bonds: Discrete Product Selection Dictated by Catalyst Ligands. *J. Am. Chem. Soc.* 1998, *120*, 9949–9950.
23. Seiser, T.; Roth, O. A.; Cramer, N. Enantioselective Synthesis of Indanols from tert-Cyclobutanols Using a Rhodium-Catalyzed C–C/C–H Activation Sequence. *Angew. Chem., Int. Ed.* 2009, *48*, 6320–6323.
24. Yu, H.; Wang, C.; Yang, Y.; Dang, Z.-M. Mechanistic Study on Rh-Catalyzed Stereoselective C–C/C–H Activation of tert-Cyclobutanols. *Chem. Eur. J.* 2014, *20*, 3839–3848.
25. Hentges, S. G.; Sharpless, K. B. Asymmetric Induction in the Reaction of Osmium Tetroxide with Olefins. *J. Am. Chem. Soc.* 1980, *102*, 4263–4265.
26. Jacobsen, E. N.; Marko, I.; Mungall, W. S.; Schroeder, G.; Sharpless, K. B. Asymmetric Dihydroxylation Via Ligand-accelerated Catalysis. *J. Am. Chem. Soc.* 1988, *110*, 1968–1970.
27. Sharpless, K. B. Searching for New Reactivity (Nobel Lecture). *Angew. Chem., Int. Ed.* 2002, *41*, 2024–2032.
28. Bolm, C.; Gerlach, A. Polymer-Supported Catalytic Asymmetric Sharpless Dihydroxylations of Olefins. *Eur. J. Org. Chem.* 1998, *1998*, 21–27.
29. Berrisford, D. J.; Bolm, C.; Sharpless, K. B. Ligand-Accelerated Catalysis. *Angew. Chem., Int. Ed.* 1995, *34*, 1059–1070.
30. Cha, J. K.; Kim, N.-S. Acyclic Stereocontrol Induced by Allylic Alkoxy Groups. Synthetic Applications of Stereoselective Dihydroxylation in Natural Product Synthesis. *Chem. Rev.* 1995, *95*, 1761–1795.
31. Kolb, H. C.; Van Nieuwenhze, M. S.; Sharpless, K. B. Catalytic Asymmetric Dihydroxylation. *Chem. Rev.* 1994, *94*, 2483–2547.
32. Sharpless, K. B.; Amberg, W.; Bennani, Y. L.; Crispino, G. A.; Hartung, J.; Jeong, K. S.; Kwong, H. L.; Morikawa, K.; Wang, Z. M. The Osmium-catalyzed Asymmetric Dihydroxylation: A New Ligand Class and a Process Improvement. *J. Org. Chem.* 1992, *57*, 2768–2771.
33. Weber, F.; Brückner, R. Asymmetric Dihydroxylation of Esters and Amides of Methacrylic, Tiglic, and Angelic Acid: No Exception to the Sharpless Mnemonic! *Eur. J. Org. Chem.* 2015, *2015*, 2428–2449.
34. Zhao, Y.; Xing, X.; Zhang, S.; Wang, D. Z. N,N-Dimethylaminobenzoates Enable Highly Enantioselective Sharpless Dihydroxylations of 1,1-Disubstituted Alkenes. *Org. Biomol. Chem.* 2014, *12*, 4314–4317.
35. Song, C. E. *Cinchona Alkaloids in Synthesis and Catalysis: Ligands, Immobilization and Organocatalysis*. Weinheim, Germany: Wiley-VCH, 2009.
36. Norrby, P. O.; Kolb, H. C.; Sharpless, K. B. Calculations on the Reaction of Ruthenium Tetroxide with Olefins Using Density Functional Theory (DFT). Implications for the Possibility of Intermediates in Osmium-catalyzed Asymmetric Dihydroxylation. *Organometallics* 1994, *13*, 344–347.
37. Göbel, T.; Sharpless, K. B. Temperature Effects in Asymmetric Dihydroxylation: Evidence for a Stepwise Mechanism. *Angew. Chem., Int. Ed.* 1993, *32*, 1329–1331.

38. Sharpless, K. B.; Teranishi, A. Y.; Backvall, J. E. Chromyl Chloride Oxidations of Olefins. Possible Role of Organometallic Intermediates in the Oxidations of Olefins by Oxo Transition Metal Species. *J. Am. Chem. Soc.* 1977, *99*, 3120–3128.

39. Corey, E. J.; Noe, M. C.; Sarshar, S. The Origin of High Enantioselectivity in the Dihydroxylation of Olefins Using Osmium Tetraoxide and Cinchona Alkaloid Catalysts. *J. Am. Chem. Soc.* 1993, *115*, 3828–3829.

40. Corey, E. J.; Jardine, P. D.; Virgil, S.; Yuen, P. W.; Connell, R. D. Enantioselective Vicinal Hydroxylation of Terminal and E-1,2-Disubstituted Olefins by a Chiral Complex of Osmium Tetroxide. An Effective Controller System and a Rational Mechanistic Model. *J. Am. Chem. Soc.* 1989, *111*, 9243–9244.

41. Deubel, D. V.; Frenking, G. [3+2] versus [2+2] Addition of Metal Oxides across CC Bonds. Reconciliation of Experiment and Theory. *Acc. Chem. Res.* 2003, *36*, 645–651.

42. DelMonte, A. J.; Haller, J.; Houk, K. N.; Sharpless, K. B.; Singleton, D. A.; Strassner, T.; Thomas, A. A. Experimental and Theoretical Kinetic Isotope Effects for Asymmetric Dihydroxylation. Evidence Supporting a Rate-Limiting "(3 + 2)" Cycloaddition. *J. Am. Chem. Soc.* 1997, *119*, 9907–9908.

43. Torrent, M.; Deng, L.; Duran, M.; Sola, M.; Ziegler, T. Density Functional Study of the [2+2]- and [2+3]-Cycloaddition Mechanisms for the Osmium-Catalyzed Dihydroxylation of Olefins. *Organometallics* 1997, *16*, 13–19.

44. Dapprich, S.; Ujaque, G.; Maseras, F.; Lledós, A.; Musaev, D. G.; Morokuma, K. Theory Does Not Support an Osmaoxetane Intermediate in the Osmium-Catalyzed Dihydroxylation of Olefins. *J. Am. Chem. Soc.* 1996, *118*, 11660–11661.

45. Pidun, U.; Boehme, C.; Frenking, G. Theory Rules Out a [2 + 2] Addition of Osmium Tetroxide to Olefins as Initial Step of the Dihydroxylation Reaction. *Angew. Chem., Int. Ed.* 1996, *35*, 2817–2820.

46. Hoi-Lun, K.; Sorato, C.; Ogino, Y.; Hou, C.; Sharpless, K. B. Preclusion of the "Second Cycle" in the Osmium-catalyzed Asymmetric Dihydroxylation of Olefins Leads to a Superior Process. *Tetrahedron Lett.* 1990, *31*, 2999–3002.

47. Norrby, P.-O.; Kolb, H. C.; Sharpless, K. B. Toward an Understanding of the High Enantioselectivity in the Osmium-Catalyzed Asymmetric Dihydroxylation. 2. A Qualitative Molecular Mechanics Approach. *J. Am. Chem. Soc.* 1994, *116*, 8470–8478.

48. Fristrup, P.; Jensen, G. H.; Andersen, M. L. N.; Tanner, D.; Norrby, P.-O. Combining Q2MM Modeling and Kinetic Studies for Refinement of the Osmium-catalyzed Asymmetric Dihydroxylation (AD) Mnemonic. *J. Organomet. Chem.* 2006, *691*, 2182–2198.

49. Fristrup, P.; Tanner, D.; Norrby, P.-O. Updating the Asymmetric Osmium-catalyzed Dihydroxylation (AD) Mnemonic: Q2MM Modeling and New Kinetic Measurements. *Chirality* 2003, *15*, 360–368.

50. Moitessier, N.; Henry, C.; Len, C.; Chapleur, Y. Toward a Computational Tool Predicting the Stereochemical Outcome of Asymmetric Reactions. 1. Application to Sharpless Asymmetric Dihydroxylation. *J. Org. Chem.* 2002, *67*, 7275–7282.

51. Moitessier, N.; Maigret, B.; Chrétien, F.; Chapleur, Y. Molecular Dynamics-based Models Explain the Unexpected Diastereoselectivity of the Sharpless Asymmetric Dihydroxylation of Allyl D-Xylosides. *Eur. J. Org. Chem.* 2000, *2000*, 995–1005.

52. Norrby, P.-O.; Rasmussen, T.; Haller, J.; Strassner, T.; Houk, K. N. Rationalizing the Stereoselectivity of Osmium Tetroxide Asymmetric Dihydroxylations with Transition State Modeling Using Quantum Mechanics-Guided Molecular Mechanics. *J. Am. Chem. Soc.* 1999, *121*, 10186–10192.

53. Ujaque, G.; Maseras, F.; Lledós, A. Theoretical Study on the Origin of Enantioselectivity in the Bis(dihydroquinidine)-3,6-pyridazine·Osmium Tetroxide-Catalyzed Dihydroxylation of Styrene. *J. Am. Chem. Soc.* 1999, *121*, 1317–1323.

54. Corey, E. J.; Noe, M. C. A Critical Analysis of the Mechanistic Basis of Enantioselectivity in the Bis-Cinchona Alkaloid Catalyzed Dihydroxylation of Olefins. *J. Am. Chem. Soc.* 1996, *118*, 11038–11053.

55. Corey, E. J.; Noe, M. C. Rigid and Highly Enantioselective Catalyst for the Dihydroxylation of Olefins Using Osmium Tetraoxide Clarifies the Origin of Enantiospecificity. *J. Am. Chem. Soc.* 1993, *115*, 12579–12580.

56. Kolb, H. C.; Andersson, P. G.; Sharpless, K. B. Toward an Understanding of the High Enantioselectivity in the Osmium-Catalyzed Asymmetric Dihydroxylation (AD). 1. Kinetics. *J. Am. Chem. Soc.* 1994, *116*, 1278–1291.

57. Krenske, E. H.; Houk, K. N. Aromatic Interactions as Control Elements in Stereoselective Organic Reactions. *Acc. Chem. Res.* 2013, *46*, 979–989.

58. Hopmann, K. H. Quantum Chemical Studies of Asymmetric Reactions: Historical Aspects and Recent Examples. *Int. J. Quantum Chem* 2015, *115*, 1232–1249.

59. Drudis-Solé, G.; Ujaque, G.; Maseras, F.; Lledós, A. A QM/MM Study of the Asymmetric Dihydroxylation of Terminal Aliphatic n-Alkenes with OsO4·(DHQD)2PYDZ: Enantioselectivity as a Function of Chain Length. *Chem.* 2005, *11*, 1017–1029.

60. Berne, B. J.; Weeks, J. D.; Zhou, R. Dewetting and Hydrophobic Interaction in Physical and Biological Systems. *Annu. Rev. Phys. Chem.* 2009, *60*, 85–103.

61. Meyer, E. E.; Rosenberg, K. J.; Israelachvili, J. Recent Progress in Understanding Hydrophobic Interactions. *Proc. Natl. Acad. Sci. U. S. A.* 2006, *103*, 15739–15746.

62. Chandler, D. Interfaces and the Driving Force of Hydrophobic Assembly. *Nature* 2005, *437*, 640–647.

63. Pratt, L. R.; Pohorille, A. Hydrophobic Effects and Modeling of Biophysical Aqueous Solution Interfaces. *Chem. Rev.* 2002, *102*, 2671–2692.

64. Xing, X.; Zhao, Y.; Xu, C.; Zhao, X.; Wang, D. Z. Electronic Helix Theory-guided Rational Design of Kinetic Resolutions by Means of the Sharpless Asymmetric Dihydroxylation Reactions. *Tetrahedron* 2012, *68*, 7288–7294.

65. Han, P.; Wang, R.; Wang, D. Z. Electronic Polarizability-based Stereochemical Model for Sharpless AD Reactions. *Tetrahedron* 2011, *67*, 8873–8878.

66. Hall, D. G. *Boronic Acids: Preparation and Applications in Organic Synthesis, Medicine and Materials.* 2nd ed. Somerset: Wiley, 2012.

67. Miyaura, N. Metal-Catalyzed Reactions of Organoboronic Acids and Esters. *Bull. Chem. Soc. Jpn.* 2008, *81*, 1535–1553.

68. Hayashi, T. Rhodium-catalyzed Asymmetric 1,4-addition of Organometallic Reagents. *Russ. Chem. Bull.* 2003, *52*, 2595–2605.

69. Berthon-Gelloz, G.; Hayashi, T. Rhodium- and Palladium-Catalyzed Asymmetric Conjugate Additions of Organoboronic Acids. In *Boronic Acids.* D. G. Hall, Ed. Weinheim, Germany: Wiley-VCH, 2010; pp. 263–313.

70. Cho, C. S.; Motofusa, S.-i.; Ohe, K.; Uemura, S.; Shim, S. C. A New Catalytic Activity of Antimony(III) Chloride in Palladium(0)-Catalyzed Conjugate Addition of Aromatics to alpha, beta-Unsaturated Ketones and Aldehydes with Sodium Tetraphenylborate and Arylboronic Acids. *J. Org. Chem.* 1995, *60*, 883–888.

71. Sakai, M.; Hayashi, H.; Miyaura, N. Rhodium-Catalyzed Conjugate Addition of Aryl- or 1-Alkenylboronic Acids to Enones. *Organometallics* 1997, *16*, 4229–4231.

72. Nishikata, T.; Yamamoto, Y.; Miyaura, N. Conjugate Addition of Aryl Boronic Acids to Enones Catalyzed by Cationic Palladium(II)–Phosphane Complexes. *Angew. Chem., Int. Ed.* 2003, *42*, 2768–2770.

73. Nishikata, T.; Yamamoto, Y.; Miyaura, N. 1,4-Addition of Arylboronic Acids and Arylsiloxanes to α,β-Unsaturated Carbonyl Compounds via Transmetalation to Dicationic Palladium(II) Complexes. *Organometallics* 2004, *23*, 4317–4324.

74. Shintani, R.; Duan, W.-L.; Hayashi, T. Rhodium-Catalyzed Asymmetric Construction of Quaternary Carbon Stereocenters: Ligand-Dependent Regiocontrol in the 1,4-Addition to Substituted Maleimides. *J. Am. Chem. Soc.* 2006, *128*, 5628–5629.

75. Hayashi, T.; Takahashi, M.; Takaya, Y.; Ogasawara, M. Catalytic Cycle of Rhodium-Catalyzed Asymmetric 1,4-Addition of Organoboronic Acids. Arylrhodium, Oxa-π-allylrhodium, and Hydroxorhodium Intermediates. *J. Am. Chem. Soc.* 2002, *124*, 5052–5058.

76. Hayashi, T. Rhodium-Catalyzed Asymmetric 1,4-Addition of Organoboronic Acids and Their Derivatives to Electron Deficient Olefins. *Synlett* 2001, 0879–0887.

77. Takaya, Y.; Ogasawara, M.; Hayashi, T.; Sakai, M.; Miyaura, N. Rhodium-Catalyzed Asymmetric 1,4-Addition of Aryl- and Alkenylboronic Acids to Enones. *J. Am. Chem. Soc.* 1998, *120*, 5579–5580.

78. Nishikata, T.; Yamamoto, Y.; Gridnev, I. D.; Miyaura, N. Enantioselective 1,4-Addition of Ar3Bi, [ArBF3]K, and ArSiF3 to Enones Catalyzed by a Dicationic Palladium(II)–Chiraphos or –Dipamp Complex. *Organometallics* 2005, *24*, 5025–5032.

79. Gini, F.; Hessen, B.; Minnaard, A. J. Palladium-Catalyzed Enantioselective Conjugate Addition of Arylboronic Acids. *Org. Lett.* 2005, *7*, 5309–5312.

80. Hayashi, T.; Senda, T.; Ogasawara, M. Rhodium-Catalyzed Asymmetric Conjugate Addition of Organoboronic Acids to Nitroalkenes. *J. Am. Chem. Soc.* 2000, *122*, 10716–10717.

81. Petri, A.; Seidelmann, O.; Eilitz, U.; Leßmann, F.; Reißmann, S.; Wendisch, V.; Gutnov, A. Pd-catalyzed Asymmetric Conjugate Addition of Arylboronic Acids to 2-nitroacrylates: A Facile Route to β2-homophenylglycines. *Tetrahedron Lett.* 2014, *55*, 267–270.

82. He, Q.; Xie, F.; Fu, G.; Quan, M.; Shen, C.; Yang, G.; Gridnev, I. D.; Zhang, W. Palladium-Catalyzed Asymmetric Addition of Arylboronic Acids to Nitrostyrenes. *Org. Lett.* 2015, *17*, 2250–2253.

83. Holder, J. C.; Zou, L.; Marziale, A. N.; Liu, P.; Lan, Y.; Gatti, M.; Kikushima, K.; Houk, K. N.; Stoltz, B. M. Mechanism and Enantioselectivity in Palladium-Catalyzed Conjugate Addition of Arylboronic Acids to β-Substituted Cyclic Enones: Insights from Computation and Experiment. *J. Am. Chem. Soc.* 2013, *135*, 14996–15007.

84. Boz, E.; Haşlak, Z. P.; Tüzün, N. Ş.; Konuklar, F. A. S. A Theoretical Study On Rh(I) Catalyzed Enantioselective Conjugate Addition Reactions of Fluoroalkylated Olefins. *Organometallics* 2014, *33*, 5111–5119.
85. Holder, J. C.; Marziale, A. N.; Gatti, M.; Mao, B.; Stoltz, B. M. Palladium-Catalyzed Asymmetric Conjugate Addition of Arylboronic Acids to Heterocyclic Acceptors. *Chem.* 2013, *19*, 74–77.
86. Boeser, C. L.; Holder, J. C.; Taylor, B. L. H.; Houk, K. N.; Stoltz, B. M.; Zare, R. N. Mechanistic Analysis of an Asymmetric Palladium-catalyzed Conjugate Addition of Arylboronic Acids to [Small Beta]-Substituted Cyclic Enones. *Chem. Sci.* 2015, *6*, 1917–1922.
87. Kikushima, K.; Holder, J. C.; Gatti, M.; Stoltz, B. M. Palladium-Catalyzed Asymmetric Conjugate Addition of Arylboronic Acids to Five-, Six-, and Seven-Membered β-Substituted Cyclic Enones: Enantioselective Construction of All-Carbon Quaternary Stereocenters. *J. Am. Chem. Soc.* 2011, *133*, 6902–6905.
88. Konno, T.; Tanaka, T.; Miyabe, T.; Morigaki, A.; Ishihara, T. A First High Enantiocontrol of an Asymmetric Tertiary Carbon Center Attached with a Fluoroalkyl Group Via Rh(I)-catalyzed Conjugate Addition Reaction. *Tetrahedron Lett.* 2008, *49*, 2106–2110.
89. Morigaki, A.; Tanaka, T.; Miyabe, T.; Ishihara, T.; Konno, T. Rhodium(i)-catalyzed 1,4-conjugate Arylation toward [Small Beta]-Fluoroalkylated Electron-deficient Alkenes: A New Entry to a Construction of a Tertiary Carbon Center Possessing a Fluoroalkyl Group. *Org. Biomol. Chem.* 2013, *11*, 586–595.
90. Lin, S.; Lu, X. Cationic Pd(II)/Bipyridine-Catalyzed Conjugate Addition of Arylboronic Acids to β,β-Disubstituted Enones: Construction of Quaternary Carbon Centers. *Org. Lett.* 2010, *12*, 2536–2539.
91. Lan, Y.; Houk, K. N. Mechanism of the Palladium-Catalyzed Addition of Arylboronic Acids to Enones: A Computational Study. *J. Org. Chem.* 2011, *76*, 4905–4909.
92. Brown, H. C.; Rao, B. C. S. A New Powerful Reducing Agent—Sodium Borohydride in the Presence of Aluminum Chloride and Other Polyvalent Metal Halides1,2. *J. Am. Chem. Soc.* 1956, *78*, 2582–2588.
93. Köster, R. Borverbindungen, I. Darstellung von Bortrialkylen und ihre Reaktionen mit Olefinen. *Justus Liebigs Ann. Chem.* 1958, *618*, 31–43.
94. Parks, D. J.; Piers, W. E.; Yap, G. P. A. Synthesis, Properties, and Hydroboration Activity of the Highly Electrophilic Borane Bis(pentafluorophenyl)borane, HB(C6F5)21. *Organometallics* 1998, *17*, 5492–5503.
95. Kabalka, G. W.; Yu, S.; Li, N.-S. Selective Hydroboration of Terminal Alkenes in the Presence of Aldehydes and Ketones. *Tetrahedron Lett.* 1997, *38*, 5455–5458.
96. Beletskaya, I.; Pelter, A. Hydroborations Catalysed by Transition Metal Complexes. *Tetrahedron* 1997, *53*, 4957–5026.
97. Tucker, C. E.; Davidson, J.; Knochel, P. Mild and Stereoselective Hydroborations of Functionalized Alkynes and Alkenes Using Pinacolborane. *J. Org. Chem.* 1992, *57*, 3482–3485.
98. Vogels, C. M.; Westcott, S. A. Recent Advances in Organic Synthesis Using Transition Metal-Catalyzed Hydroborations. *Curr. Org. Chem.* 2005, *9*, 687–699.
99. Smith, M. R. Advances in Metal Boryl and Metal-Mediated B—X Activation Chemistry. In *Progress in Inorganic Chemistry*. John Wiley & Sons, 2007; pp. 505–567.

100. Evans, D. A.; Fu, G. C.; Hoveyda, A. H. Rhodium(I)- and Iridium(I)-Catalyzed Hydroboration Reactions: Scope and Synthetic Applications. *J. Am. Chem. Soc.* 1992, *114*, 6671–6679.

101. Ramachandran, P. V.; Jennings, M. P.; Brown, H. C. Critical Role of Catalysts and Boranes for Controlling the Regioselectivity in the Rhodium-Catalyzed Hydroboration of Fluoroolefins. *Org. Lett.* 1999, *1*, 1399–1402.

102. Segarra, A. M.; Claver, C.; Fernandez, E. New Insights on the Asymmetric Hydroboration of Perfluoroalkenes. *Chem. Commun.* 2004, 464–465.

103. Segarra, A. M.; Daura-Oller, E.; Claver, C.; Poblet, J. M.; Bo, C.; Fernández, E. In Quest of Factors That Control the Enantioselective Catalytic Markovnikov Hydroboration/Oxidation of Vinylarenes. *Chem.* 2004, *10*, 6456–6467.

104. Crudden, C. M.; Hleba, Y. B.; Chen, A. C. Regio- and Enantiocontrol in the Room-Temperature Hydroboration of Vinyl Arenes with Pinacol Borane. *J. Am. Chem. Soc.* 2004, *126*, 9200–9201.

105. Luna, A. P.; Bonin, M.; Micouin, L.; Husson, H.-P. Reversal of Enantioselectivity in the Asymmetric Rhodium- versus Iridium-Catalyzed Hydroboration of Meso Substrates. *J. Am. Chem. Soc.* 2002, *124*, 12098–12099.

106. Wilczynski, R.; Sneddon, L. G. Transition-metal-promoted Reactions of Boron Hydrides. 2. Synthesis and Thermolysis Reactions of Alkenyl-pentaboranes. New Synthesis of Monocarbon Carboranes. *Inorg. Chem.* 1981, *20*, 3955–3962.

107. Wilczynski, R.; Sneddon, L. G. Transition-metal-promoted Reactions of Boron Hydrides. 3. (R2C2)Co2(CO)6-catalyzed Reactions of Alkynes and Small Carboranes: Synthesis of Boron-substituted Alkenylcarboranes. *Inorg. Chem.* 1982, *21*, 506–514.

108. Männig, D.; Nöth, H. Catalytic Hydroboration with Rhodium Complexes. *Angew. Chem. Int. Ed.* 1985, *24*, 878–879.

109. Sato, M.; Miyaura, N.; Suzuki, A. Rhodium(i)-catalyzed Asymmetric Hydroboration of Alkenes with 1,3,2-benzodioxaborole. *Tetrahedron Lett.* 1990, *31*, 231–234.

110. Evans, D. A.; Fu, G. C.; Hoveyda, A. H. Rhodium(I)-catalyzed Hydroboration of Olefins. The Documentation of Regio- and Stereochemical Control in Cyclic and Acyclic Systems. *J. Am. Chem. Soc.* 1988, *110*, 6917–6918.

111. Burgess, K.; Ohlmeyer, M. J. Enantioselective Hydroboration Mediated by Homochiral Rhodium Catalysts. *J. Org. Chem.* 1988, *53*, 5178–5179.

112. Hayashi, T.; Matsumoto, Y.; Ito, Y. Catalytic Asymmetric Hydroboration of Styrenes. *J. Am. Chem. Soc.* 1989, *111*, 3426–3428.

113. Brown, J. M.; Lloyd-Jones, G. C. Catalytic Asymmetric Hydroboration with Oxazaborolidines. *Tetrahedron Asymmetry* 1990, *1*, 869–872.

114. Zeng, X. Recent Advances in Catalytic Sequential Reactions Involving Hydroelement Addition to Carbon–Carbon Multiple Bonds. *Chem. Rev.* 2013, *113*, 6864–6900.

115. Carroll, A.-M.; O'Sullivan, T. P.; Guiry, P. J. The Development of Enantioselective Rhodium-Catalysed Hydroboration of Olefins. *Adv. Synth. Catal.* 2005, *347*, 609–631.

116. Crudden, C. M.; Edwards, D. Catalytic Asymmetric Hydroboration: Recent Advances and Applications in Carbon–Carbon Bond-Forming Reactions. *Eur. J. Org. Chem.* 2003, *2003*, 4695–4712.

117. Fernandez, E.; Maeda, K.; Hooper, M. W.; Brown, J. M. Catalytic Asymmetric Hydroboration/Amination and Alkylamination with Rhodium Complexes of 1,1'-(2-Diarylphosphino-1-naphthyl)isoquinoline. *Chem.* 2000, *6*, 1840–1846.
118. Ren, L.; Crudden, C. M. Homologations of Boronate Esters: The First Observation of Sequential Insertions. *Chem. Commun.* 2000, 721–722.
119. Chen, A.; Ren, L.; Crudden, M. C. Catalytic Asymmetric Carbon–Carbon Bond Forming Reactions: Preparation of Optically Enriched 2-aryl Propionic Acids by a Catalytic Asymmetric Hydroboration-Homologation Sequence [Dagger]. *Chem. Commun.* 1999, 611–612.
120. Fernandez, E.; Hooper, M.; Knight, F.; Brown, J. Catalytic Asymmetric Hydroboration-Amination. *Chem. Commun.* 1997, 173–174.
121. Daura-Oller, E.; Segarra, A. M.; Poblet, J. M.; Claver, C.; Fernández, E.; Bo, C. On the Origin of Regio- and Stereoselectivity in the Rhodium-Catalyzed Vinylarenes Hydroboration Reaction. *J. Org. Chem.* 2004, *69*, 2669–2680.
122. Hayashi, T.; Matsumoto, Y.; Ito, Y. Asymmetric Hydroboration of Styrenes Catalyzed by Cationic Chiral Phosphine-Rhodium (I) Complexes. *Tetrahedron Asymmetry* 1991, *2*, 601–612.
123. Doucet, H.; Fernandez, E.; Layzell, T. P.; Brown, J. M. The Scope of Catalytic Asymmetric Hydroboration/Oxidation with Rhodium Complexes of 1,1'-(2-Diarylphosphino-1-Naphthyl)Isoquinolines. *Chem.* 1999, *5*, 1320–1330.
124. Kwong, F. Y.; Yang, Q.; Mak, T. C. W.; Chan, A. S. C.; Chan, K. S. A New Atropisomeric P,N Ligand for Rhodium-Catalyzed Asymmetric Hydroboration. *J. Org. Chem.* 2002, *67*, 2769–2777.
125. Noyori, R.; Ohkuma, T. Asymmetric Catalysis by Architectural and Functional Molecular Engineering: Practical Chemo- and Stereoselective Hydrogenation of Ketones. *Angew. Chem., Int. Ed.* 2001, *40*, 40–73.
126. Widauer, C.; Grützmacher, H.; Ziegler, T. Comparative Density Functional Study of Associative and Dissociative Mechanisms in the Rhodium(I)-Catalyzed Olefin Hydroboration Reactions. *Organometallics* 2000, *19*, 2097–2107.
127. Soai, K.; Shibata, T.; Morioka, H.; Shoji, K. Asymmetric Autocatalysis and Amplification of Enantiomeric Excess of a Chiral Molecule. *Nature* 1995, *378*, 767.
128. Soai, K.; Shibata, T.; Sato, I. Enantioselective Automultiplication of Chiral Molecules by Asymmetric Autocatalysis. *Acc. Chem. Res.* 2000, *33*, 382.
129. Soai, K.; Kawasaki, T. Discovery of Asymmetric Autocatalysis with Amplification of Chirality and Its Implications in Chiral Homogeneity of Biomolecules. *Chirality* 2006, *18*, 469.
130. Soai, K.; Kawasaki, T. Asymmetric Autocatalysis with Amplification of Chirality. *Top. Curr. Chem.* 2008, *271–274*, 1.
131. Kawasaki, T.; Soai, K. Asymmetric Induction Arising from Enantiomerically Enriched Carbon-13 Isotopomers and Highly Selective Chiral Discrimination by Asymmetric Autocatalysis. *Bull. Chem. Soc. Jpn.* 2011, *84*, 879–892.
132. Soai, K.; Shibata, T.; Kowata, Y. Asymmetric Synthesis of Enantioenriched Alkanol by Spontaneous Asymmetric Synthesis. *Jap. Kokai Tokkyo Koho* 1997, 9-268179.
133. Singleton, D. A.; Vo, L. K. Enantioselective Synthesis without Discrete Optically Active Additives. *J. Am. Chem. Soc.* 2002, *124*, 10010–10011.

134. Soai, K.; Kawasaki, T.; Matsumoto, A. Asymmetric Autocatalysis of Pyrimidyl Alkanol and Its Application to the Study on the Origin of Homochirality. *Acc. Chem. Res.* 2014, *47*, 3643–3654.
135. Gridnev, I. D.; Serafimov, J. M.; Quiney, H.; Brown, J. M. Reflections on Spontaneous Asymmetric Synthesis by Amplifying Autocatalysis. *Org. Biomol. Chem.* 2003, *1*, 3811.
136. Singleton, D. A.; Vo, L. K. A Few Molecules Can Control the Enantiomeric Outcome. Evidence Supporting Absolute Asymmetric Synthesis Using the Soai Asymmetric Autocatalysis. *Org. Lett.* 2003, *5*, 4337.
137. Barabas, B.; Caglioti, L.; Zucchi, C.; Maioli, M.; Gal, E.; Micskei, K.; Palyi, G. Violation of Distribution Symmetry in Statistical Evaluation of Absolute Enantioselective Synthesis. *J. Phys. Chem. B* 2007, *111*, 11506–11510.
138. Micskei, K.; Rabai, G.; Gal, E.; Caglioti, L.; Palyi, G. Oscillatory Symmetry Breaking in the Soai Reaction. *J. Phys. Chem. B* 2008, *112*, 9196–9200.
139. Suzuki, K; Hatase, K.; Nishiyama, D.; Kawasaki, T.; Soai, K. Spontaneous Absolute Asymmetric Synthesis Promoted by Achiral Amines in Conjunction with Asymmetric Autocatalysis. *J. Syst. Chem.* 2010, *1*, 5.
140. Frank, F. C. Spontaneous Asymmetric Synthesis. *Biochim. Biophys. Acta* 1953, *11*, 459–463.
141. Avetisov, V.; Goldanskii, V. I. Minor Symmetry Breaking at the Molecular Level. *Proc. Natl. Acad. Sci. U. S. A.* 1996, *93*, 11435–11442.
142. Goldanskii, V. I.; Kuzmin, V. V. Z. Spontaneous Mirror Symmetry Breaking in Nature and the Origin of Life. *Phys. Chem. (Leipzig)* 1988, *269*, 216–274.
143. Mislow, K. Absolute Asymmetric Synthesis: A Commentary. *Collect. Czech. Chem. Commun.* 2003, *68*, 849–864.
144. Caglioti, L.; Hajdu, C.; Holczknecht, O.; Zékány, L.; Zucchi, C.; Micskei, K.; Pályi, G. The Concept of Racemates and the Soai Reaction. *Viva Origino* 2006, *34*, 62–80.
145. Buhse, T. A Tentative Kinetic Model for Chiral Amplification in Autocatalytic Alkylzinc Additions. *Tetrahedron Asymmetry* 2003, *14*, 1055–1061.
146. Girard, G.; Kagan, H. B. Nonlinear Effects in Asymmetric Synthesis and Stereoselective Reactions: Ten Years of Investigation. *Anew. Chem. Int. Ed.* 1998, *37*, 2922–2959.
147. Blackmond, D. G. Mechanistic Study of the Soai Autocatalytic Reaction Informed by Kinetic Analysis. *Tetrahedron Asymmetry* 2006, *17*, 584–589.
148. Islas, J. R.; Buhse, T. Kinetic Modelling of Chiral Amplification and Enantioselectivity Reversal in Asymmetric Reactions. *J. Mex. Chem. Soc.* 2007, *51*, 117–121.
149. Ribo, J. M.; Hochberg, D. Stability of Racemic and Chiral Steady States in Open and Closed Chemical Systems. *Physics Letters A* 2008, *373*, 111–122.
150. Crusats, J.; Hochberg, D.; Moyano, A.; Ribo, J. M. Frank Model and Spontaneous Emergence of Chirality in Closed Systems. *Chem. Phys. Chem* 2009, *10*, 2123–2131.
151. Micheau, J.-C.; Cruz, J.-M.; Coudret, C.; Buhse, T. An Autocatalytic Cycle Model of Asymmetric Amplification and Mirror-Symmetry Breaking in the Soai Reaction. *Chem. Phys. Chem* 2010, *11*, 3417–3419.
152. Blackmond, D. G. Kinetic Aspects of Non-Linear Effects in Asymmetric Synthesis, Catalysis, and Autocatalysis. *Tetrahedron Asymmetry* 2010, *21*, 1630–1634.

153. Doka, E.; Lente, G. Mechanism-Based Chemical Understanding of Chiral Symmetry Breaking in the Soai Reaction. A Combined Probabilistic and Deterministic Description of Chemical Reactions. *J. Am. Chem. Soc.* 2011, *133*, 17878–17881.

154. Gridnev, I. D.; Vorob'ev, A. K.; Raskatov, E. A. Role of Oligomerization in Soai Reaction: Structures, Energies, Possible Reactivity. In: *The Soai Reaction and Related Topic*. G. Pályi. C. Zucchi, L. Caglioti, Eds. Modena, Italy: Artestampa, 2012, pp. 79–122.

155. Gridnev, I. D.; Vorob'ev, A. K. Quantification of Sophisticated Equilibria in the Reaction Pool and Amplifying Catalytic Cycle of Soai Reaction. *ACS Catal.* 2015, *2*, 2137–2149.

156. Gridnev, I. D.; Serafimov, J. M.; Brown, J. M. Solution Structure and Reagent Binding of the Zinc Alcoxide Catalyst in Soai's Asymmetric Autocatalytic Reaction. *Angew. Chem. Int. Ed. Engl.* 2004, *43*, 4884–4487.

157. Klankermayer, J.; Gridnev, I. D.; Brown, J. M. Role of the Isopropyl Group in Asymmetric Autocatalytic Zinc Alkylations. *Chem. Comm.* 2007, 3151–3153.

158. Brown, J. M.; Gridnev, I. D.; Klankermayer, J. Asymmetric Autocatalysis with Organozinc Complexes; Elucidation of the Reaction Pathway. *Top. Curr. Chem.* 2008, *284*, 35–65.

159. Gridnev, I. D.; Brown, J. M. Asymmetric Autocatalysis: Novel Structures, Novel Mechanism? *Proc. Natl. Acad. Sci. U. S. A.* 2004, *101*, 5727–5731.

160. Gridnev, I. D. Chiral Symmetry Breaking in Chiral Crystallization and Soai Autocatalytic Reaction. *Chem. Lett.* 2006, 148–153.

161. Blackmond, D. G.; McMillan, C. R.; Ramdeehul, S.; Schorm, A.; Brown, J. M. Origins of Asymmetric Amplification in Autocatalytic Alkylzinc Additions. *J. Am. Chem. Soc.* 2001, *123*, 10103–10105.

162. Schiaffino, L.; Ercolani, G. Unraveling the Mechanism of the Soai Asymmetric Autocatalytic Reaction by First-Principles Calculations: Induction and Amplification of Chirality by Self-Assembly of Hexamolecular Complexes. *Angew. Chem. Int. Ed.* 2008, *47*, 6832–6835.

163. Schiaffino, L.; Ercolani, G. Amplification of Chirality and Enantioselectivity in the Asymmetric Autocatalytic Soai Reaction. *Chem. Phys. Chem.* 2009, *10*, 2508–2515.

164. Schiaffino, L.; Ercolani, G. Mechanism of the Asymmetric Autocatalytic Soai Reaction Studied by Density Functional Theory. *Chem. Eur. J.* 2010, *16*, 3147–3156.

165. Ercolani, G.; Schiaffino, L. Putting the Mechanism of the Soai Reaction to the Test: DFT Study of the Role of Aldehyde and Dialkylzinc Structure. *J. Org. Chem.* 2011, *76*, 2619–2626.

166. Gehring, T.; Quaranta, M.; Odell, B.; Blackmond, D. G.; Brown, J. M. Observation of a Transient Intermediate in soai's Asymmetric Autocatalysis: Insights from [1]H Turnover in Real Time. *Angew. Chem. Int. Ed.* 2012, *51*, 9539–9542.

167. Gridnev, I. D.; Vorobiev, A. K. On the Origin and Structure of the Recently Observed Acetal in Soai Reaction. *Bull. Chem. Soc. Jpn.*, 2015, *88*, 333–340.

chapter three

Mechanism of enantioselection in organocatalytic reactions

3.1 Phosphoric acids

3.1.1 Asymmetric allylboration

Originally, enantioselective allylboration was developed using chiral allylbo-ranes and allyl boronates.[1-5] These reactions require multistep preparations of chiral reagents that are used in stoichiometric amounts, and are therefore impractical. Recently, catalytic asymmetric allylborations were developed. These reactions can apply either chiral Lewis bases[6-10] or Brønsted acids[11-14] as the catalysts. In particular, chiral BINOL-phosphoric acids were demonstrated to provide high optical yields in the enantioselective allylboration reaction between allylboronate 1 and aldehydes.[15] For example, the catalytic asymmetric allylboration of benzaldehyde 2 proceeded quantitatively yielding the corresponding homoallyl alcohol 3 with 98% ee (Scheme 3.1).

The mechanism of enantioselection was examined computationally by two groups.[16-18] In both studies, it was suggested that the catalyst does not change dramatically the six-membered transition state of the allylboration reaction (Figure 3.1), but decreases its activation barrier via specific coordination of the chiral phosphoric acid to one of the oxygen atoms of the boronate ring.[16-18]

Each type of coordination of the chiral phosphoric acid can result in either R- or S-allylation product depending on the prochiral plane of benzaldehyde, which is facing the allyl group of the boronate. This gives four possible transition states **TS1R**, **TS2R**, **TS1S**, and **TS2S** (Scheme 3.2, Figure 3.2).[18]

Computed relative stabilities of the transition states (Figure 3.2) are in accord with high R-enantioselectivity of the reaction. The significant instability of the **TS1S** compared to **TS1R** is explained by unavoidable close contacts between the isopropyl substituents of the catalyst and the methyl groups of the allylboronate (Figure 3.2). Explanation of a much smaller energy difference between **TS2S** and **TS2R** is much less straightforward.

The computational results testify that the level of enantioselection in this reaction is determined via the difference in stabilities of **TS1R** and **TS2S**, that is, of the transition states with significantly different structures. Thus, the **TS1R** is stabilized by the weak hydrogen bond of the

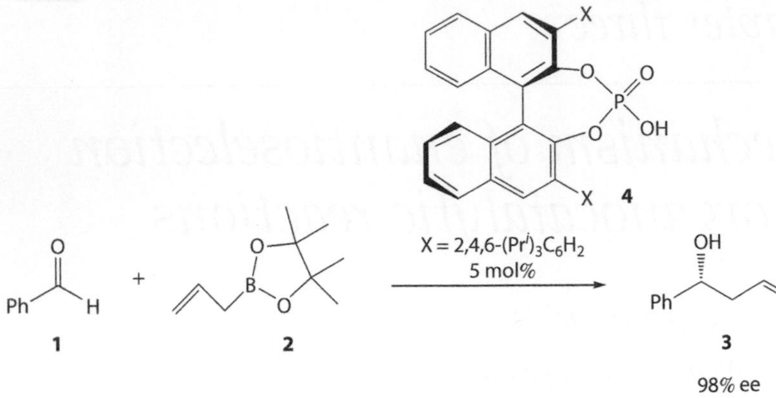

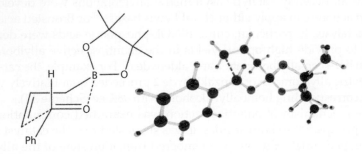

Scheme 3.1 Asymmetric catalytic allylboration. (Data from Jain, P. and Antilla, J. C., *J. Am. Chem. Soc.*, 132, 11884–11886, 2010.)

Figure 3.1 Transition state of noncatalytic allylboration. (Data from Wang, H. et al., *J. Org. Chem.*, 78, 1208–1215, 2013.)

aldehyde proton, whereas in the **TS2S** the weak stabilizing interactions of the P=O unit with the *o*-proton of the benzaldehyde aromatic ring and one of the isopropyl groups of the catalyst are observed. Different structures of **TS1R** and **TS2S** make direct analysis of the factors responsible for the high enantioselection difficult. Nevertheless, it was possible to conclude that the pinacol boronate methyls have an important role in the stereoselection, and these groups could be altered to influence stereoselectivities.[18]

3.1.2 Kinetic resolution in Robinson-type cyclization

Cyclization product **6** was selectively obtained from the racemic precursor **5** as a result of the kinetic resolution in Robinson-type cyclization (Scheme 3.3).[19]

Computational study showed that the C–C bond formation can be selectively facilitated via bifunctional coordination of the catalyst (Figure 3.3).

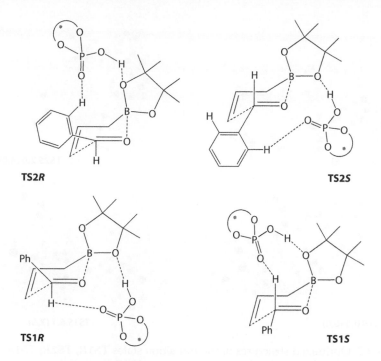

TS2*R*

TS2*S*

TS1*R*

TS1*S*

Scheme 3.2 Schematic drawings of the transition states in asymmetric catalytic allylboration.

In the case of the *R*-enantiomer, the coordinated substrate avoids any close contacts with the substituents of **4** in the **TS3**. On the other hand, the **TS4** leading to the formation of another enantiomer of **6**, the bifunctional coordination of the substrate determines its orientation in which both sides of the molecule interfere with the bulky substituents of the catalyst (Figure 3.3).

It should be noted that since the optimizations were made with ONIOM approach computing most of the aromatic rings at HF/3-21 level of theory, the important weak intramolecular interactions could be neglected in this study.

3.1.3 *Friedel–Crafts alkylation of indoles with nitroalkenes*

Indole **7** was enantioselectively alkylated by nitrostyrene **8** applying chiral phosphoric acid catalysis (Scheme 3.4). The highest enantioselectivity of the product **9** was obtained when the bis-TMS substituted phosphoric acid was applied.[19]

Computational study of the origin of enantioselection[20,21] revealed a bifunctional mechanism of the substrate activation that also followed

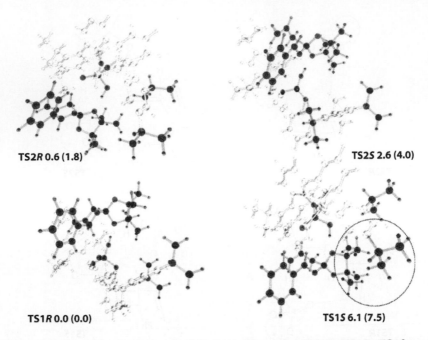

TS2R 0.6 (1.8)   **TS2S 2.6 (4.0)**

TS1R 0.0 (0.0)   **TS1S 6.1 (7.5)**

Figure 3.2 Optimized structures of the transition states **TS1R, TS2R, TS1S**, and **TS2S**. In full brightness are shown the reactants, HPO_4 unit of the catalyst and the isopropyl substituents interfering with the allylboronate in the **TS1S**. Relative energies in kcal/mol are shown for the B3LYP/6-31G* optimizations in gas phase, the values in brackets are results of energy calculations with B3LYP-D3. (Adapted with permission from Wang, H. et al., *J. Org. Chem.*, 78, 1208–1215. Copyright 2013 American Chemical Society.)

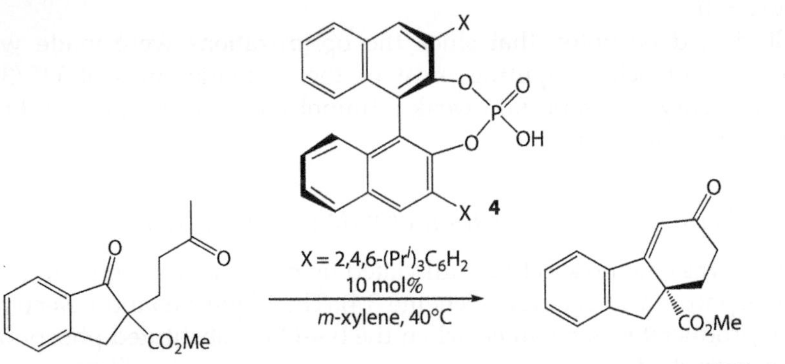

$X = 2,4,6\text{-}(Pr^i)_3C_6H_2$
10 mol%
m-xylene, 40°C

5 (*rac*)   **6**   92% ee

Scheme 3.3 Kinetic resolution catalyzed by chiral BINOL-based phosphoric acids.

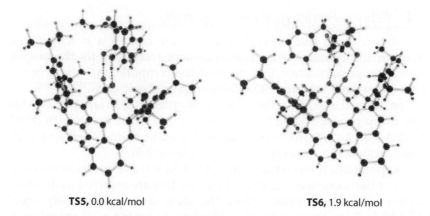

TS5, 0.0 kcal/mol   **TS6,** 1.9 kcal/mol

Figure 3.3 The 3D structures of **TS3** and **TS4**. (From Yamanaka, M. et al.: Kinetic Resolution in Chiral Phosphoric Acid Catalyzed Aldol Reactions: Enantioselective Robinson-Type Annulation Reactions. *Eur. J. Org. Chem.* 2012. 24, 4508–4514. Copyright Wiley-VCH Verlag GmbH & Co. KGaA. Adapted with permission.)

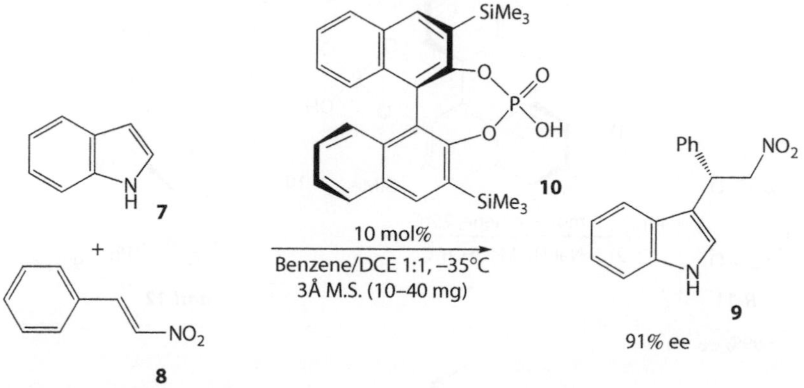

Scheme 3.4 Friedel–Crafts alkylation of indole with nitrostyrene.

from the poor results observed for a N-methylated indole.[19] Four transition states, **TS5–TS8**, possible for a bifunctional activation were optimized on the ONIOM (B3LYP/6-31G*:HF/3-21G) level of theory. The preference for the formation of the experimentally observed *S*-enantiomer for 2.7 kcal/mol was obtained in these computations. Moreover, the effects of the substituents in the catalyst and methylation of indole were computationally reproduced. However, significant number of possible weak interactions between substrates and/or the substrates and the catalyst neglected in this study demand some caution in accepting the suggested stereoregulating factors. The structures of **TS5–TS8** can be found in the CD.

3.1.4 Petasis–Ferrier-type rearrangement

The catalytic cycle and stereochemical preferences of a stoichiometric variant of the Petasis–Ferrier rearrangement were recently thoroughly studied computationally.[22] This enabled the computational study of the reaction catalyzed by chiral BINOL-based phosphoric acids.[23]

The reaction of (*R*)-**11** in the presence of catalyst (*R*)-**10** afforded *anti*-**12a** as the major product, while the reaction of (*S*)-**11** in the presence of (*R*)-**13** afforded *syn*-**12** as the major product. Both reactions proceeded with high chirality transfer in a retentive manner (Scheme 3.5).[23]

Computations showed that during the whole transformation the bond lengths of two oxygen atoms of the PO_4 unit that are involved in the coordination with the substrate remain the same and approximately equal (1.47–1.50 Å). Thus, the phosphoric acid releases a proton as an anionic conjugate base, and the resultant negative charge is delocalized over the phosphoric acid moiety. As a result, the interaction between the catalyst and the

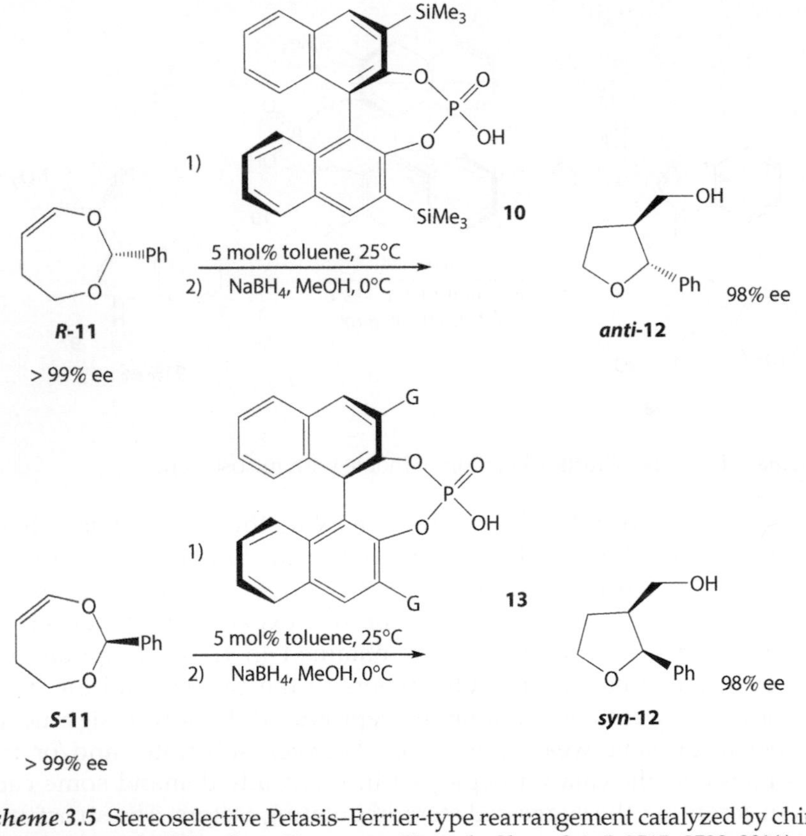

Scheme 3.5 Stereoselective Petasis–Ferrier-type rearrangement catalyzed by chiral phosphoric acids. (Data from Kanomata, K. et al., *Chem. Sci.*, 5, 3515–3523, 2014.)

substrate is better described as a complex between the negatively charged conjugate base of the catalyst and the positively charged (protonated) substrate, which is stabilized by hydrogen-bonding interactions between the enol moiety of the substrate and the phosphate moiety of the catalyst.

The relative stabilities of the diastereomeric transition states are determined by the number of the stabilizing hydrogen bonds. Thus, the **TS9** which affords *anti*-**12**, is stabilized by one O–H...O and two C–H...O hydrogen bonds. In contrast, the **TS10** leading to the *syn*-**12** is stabilized by only two hydrogen bonds: one O–H...O and one C–H...O hydrogen bond with *ortho*-hydrogen of the phenyl substituent of the substrate, since the C–H bond of the oxocarbenium moiety is oriented away from the catalyst and unable to form the C–H/O hydrogen bond (Figure 3.4).[23]

Although in the case of the rearrangement of **S-11** catalyzed by **13**, the number of the stabilizing hydrogen bonds in the **TS11** (leading to the *anti*-**12**) is also larger than that in the **TS12** (leading to the *syn*-**12**), the latter transition state is more stable due to the existence of a π–π stacking interaction with the antryl substituent in **13** (Figure 3.5).[23]

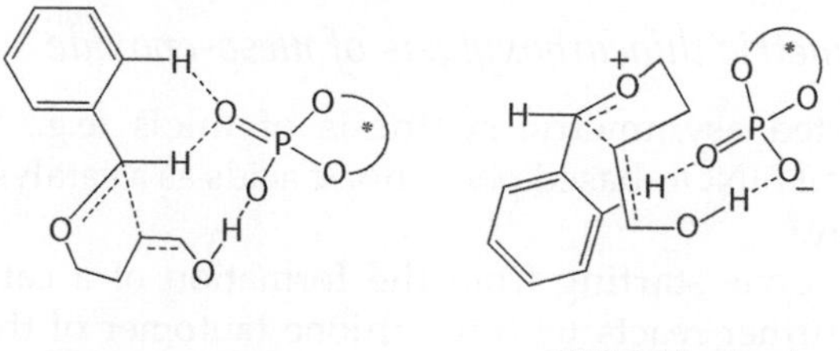

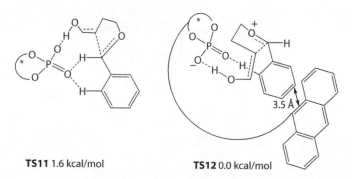

TS9 0.0 kcal/mol   **TS10** 3.6 kcal/mol

Figure 3.4 Schematic representation of the hydrogen bonding in **TS11** and **TS12**. (Reproduced from Kanomata, K. et al., *Chem. Sci.*, 5, 3515–3523, 2014. With permission from The Royal Society of Chemistry.)

TS11 1.6 kcal/mol   **TS12** 0.0 kcal/mol

Figure 3.5 Schematic representation of the hydrogen bonding in **TS11** and **TS12**. (Kanomata, K. et al., *Chem. Sci.*, 5, 3515–3523, 2014. Adapted with permission from The Royal Society of Chemistry.)

Thus, the stereoselection in the Petasis–Ferrier-type rearrangement catalyzed by chiral phosphoric acids is based according to this study on a subtle interplay of the nonconventional weak interactions. The importance of the aldehyde C–H...O hydrogen bonding has been earlier suggested in various transformations catalyzed by chiral phosphoric acids.[24–26]

3.1.5 Enantioselective indole aza-Claisen rearrangement

Reasonably high optical yield (90% ee) was recently obtained in an aza-Claisen rearrangement of the substituted indole **14** catalyzed by the chiral phosphoric acid **13** (Scheme 3.6).[27]

The transition state **TS13** leading to the major product **15** was optimized on the B3LYP/6-31G(d) level of theory (Figure 3.6). An edge-to-face CH–π interaction between the Ph group of the substrate and the 9-anthracene group as well as a CH–O interaction between this phenyl group and the phosphate is present.[27]

3.1.6 Asymmetric thiocarboxylysis of meso-epoxide

Recently reported asymmetric synthesis of thiols (e.g., Scheme 3.7)[28] employing chiral BINOL-based phosphoric acids as a catalyst was studied computationally.[29]

A catalytic cycle starting from the formation of a catalyst–epoxide complex that further reacts with the thione tautomer of the thiobenzoic acid **17** (Scheme 3.7) was computed. Full molecule of the catalyst was used in the PCM-M06-2X/6-311+G**//M06-2X/6-31G* computations.

The transition states **TS14** and **TS15** leading to opposite enantiomers of the product were found to differ in free energies for 1.5 kcal/mol in

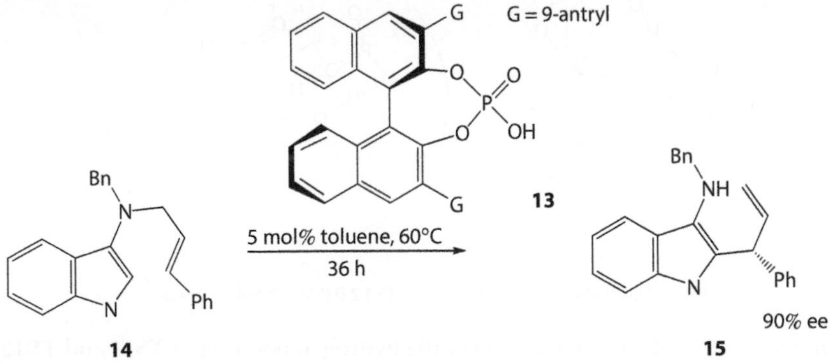

Scheme 3.6 Enantioselective indole aza-Claisen rearrangement.

TS13

Figure 3.6 Optimized structure of **TS13**. (Adapted with permission from Maity, P. et al., *J. Am. Chem. Soc.*, 135, 16380–16383. Copyright 2013 American Chemical Society.)

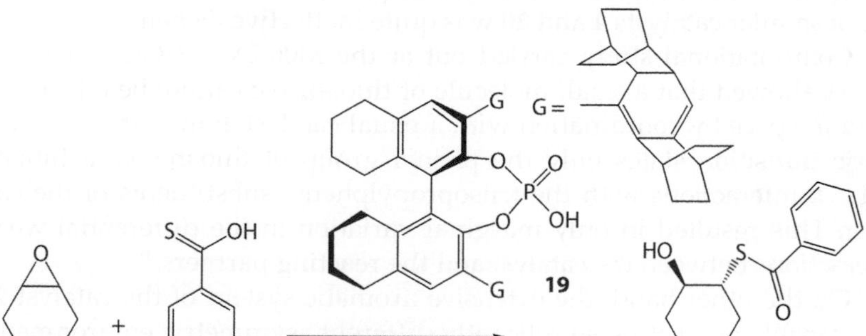

Scheme 3.7 Hydrogenated BINOL-catalyzed thiocarboxylysis of mesoepoxide to generate β-hydroxy thioesters. (Data from Ajitha, M. and Huang, K.-W., *Org. Biomol. Chem.*, 13, 10981–1985, 2015.)

good agreement with the experimental observation. However, it was difficult to rationalize the origin in this difference in stabilities, since both transition states have the same number of stabilizing interactions and are free from special strain. It was concluded that "the reactants inside the chiral cavity are more distorted in the case of **TS15** than that of **TS14**."[29]

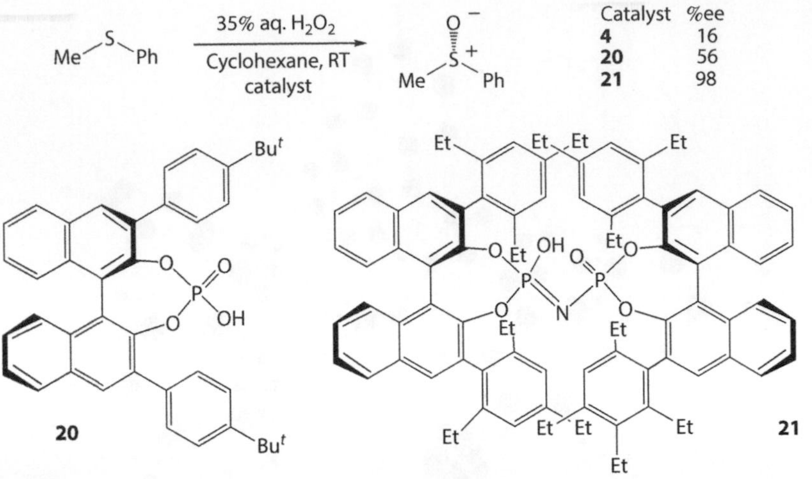

Scheme 3.8 Dependence of the optical yield in asymmetric sulfoxidation on the size of the catalyst. (Data from Jindal, G. and Sunoj, R. B., *Angew. Chem. Int. Ed.*, 53, 4432–4436, 2014.)

3.1.7 Asymmetric sulfoxidation reaction

It was reported recently that a highly enantioselective sulfoxidation can be carried out in the presence of imidodiphosphonic acid **21**, whereas the use of smaller catalysts **4** and **20** was quite ineffective (Scheme 3.8).[30]

Computational study carried out at the M06-2X/6-31G(d,p) level of theory showed that a small molecule of thioanisole cannot be effectively fixed in space by coordination with a usual catalyst **4**: in both diastereo-meric transition states only the phenyl group of thioanisole exhibited CH...π interactions with the triisopropylphenyl substituents of the cat-alyst. This resulted in only marginal variation in the differential weak interactions between the catalyst and the reacting partners.[31]

On the other hand, the extensive aromatic system of the catalyst **21** was capable of creating significantly different asymmetric environments for the alternative modes of thioanisole coordination (Figure 3.7).

As can be seen from Figure 3.7, a much wider network of CH...π and CH...O interactions present in the **TS16** compared to that in the **TS17** resulted in significant energy difference for these two transition states.

3.2 Cinchona alkaloids

After initial reports that natural alkaloids like quinidine (**QD-1**) can be applied as bifunctional chiral organocatalysts,[32] several synthetically modified catalysts were prepared (Scheme 3.9) for the enhancement of their catalytic activity and enantioselectivity.[33,34]

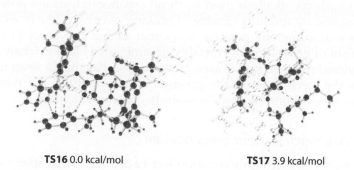

TS16 0.0 kcal/mol   **TS17** 3.9 kcal/mol

Figure 3.7 Stabilizing CH...π (blue) and CH...O (red) interactions found in the transition states for the sulfoxidation of thioanisole catalyzed by imidodiphosphonic acid **21**. (From Jindal, G. and Sunoj, R. B.: Axially Chiral Imidodiphosphoric Acid Catalyst for Asymmetric Sulfoxidation Reaction: Insights on Asymmetric Induction. *Angew. Chem. Int. Ed.* 2014. 53, 4432–4436. Copyright Wiley-VCH Verlag GmbH & Co. KGaA. Adapted with permission.)

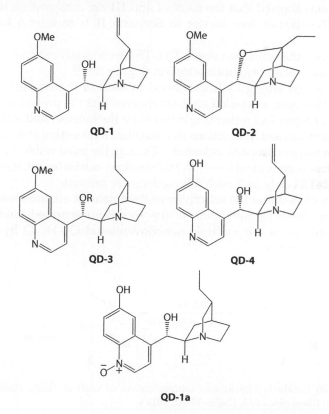

Scheme 3.9 Cinchona alkaloids used as organocatalysts.

All Cinchona alkaloids used as chiral organocatalysts are polyfunctional organic compounds capable of making reactive adducts with the substrates in numerous possible ways that makes the analysis of the mechanism of generation of chirtality an ambiguous and extensive task that hindered research in this area. Nevertheless, recently in several publications the mechanism of the generation of chirality in the reactions catalyzed by cinchona alkaloids was studied computationally.

3.2.1 Asymmetric olefin isomerization

In 2011, a general and highly enantioselective olefin isomerization via biomimetic proton-transfer catalysis with a novel cinchona alkaloid derived catalyst **QD1a** was reported (Scheme 3.10).[34]

Three catalytic cycles differing in the mode of the substrate activation were computed (Scheme 3.11). In all three scenarios, the reaction was considered to proceed through initial α-deprotonation of the substrate and the subsequent stereoselective and rate-determining γ-protonation.[34] Computations showed that the routes **I** and **III** are competitive, whereas the overall activation-free barrier in Scenario **II** is at least 6 kcal/mol higher.

Therefore, the transition states **TS1–TS4** were analyzed to clarify the intrinsic mechanism of enantioselection (Figure 3.8).[35]

There is one strong hydrogen bond in each of the transition states **TS1–TS4**. However, a detailed analysis showed that the preference for the formation of *S*-product rather originates from the number and strength of the weak noncovalent interactions that stabilize a transition state and help to keep the reagent properly oriented.[35] Thus, in the most stable among the four **TS1**, five noncovalent C—H...O interactions can be found, whereas in any of **TS2–TS4** only four such interactions are present.

It is concluded that the selectivity of the cinchona alkaloids catalyzed asymmetric olefin isomerization of β,γ- to α,β-unsaturated butenolides arises mainly from the multiple nonconventional C—H...O hydrogen-bonding interactions.

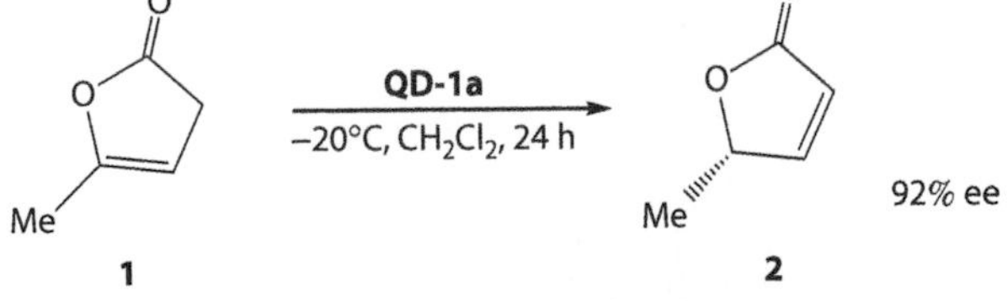

Scheme 3.10 Catalytic biomimetic isomerization of olefins. (Data from Wu, Y. et al., *J. Am. Chem. Soc.*, 133, 12458–12461, 2011.)

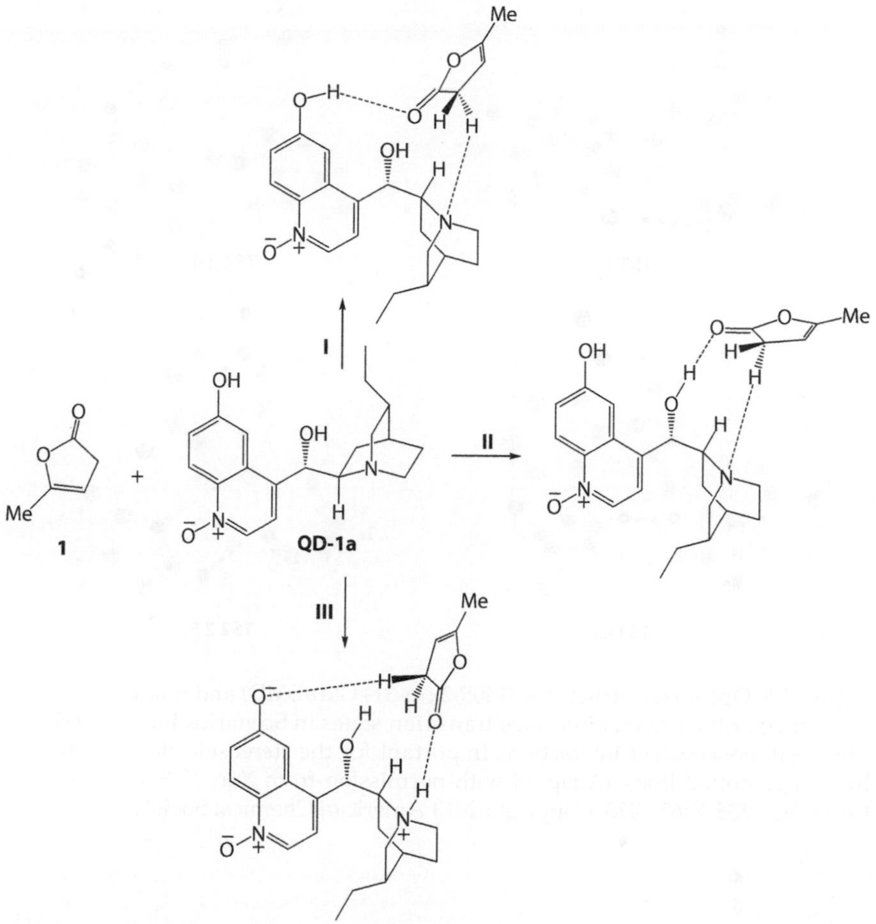

Scheme 3.11 Three routes computed for the isomerization of **1** catalyzed by cinchona alkaloid **QD-1a**. (Data from Xue, X. S. et al., *J. Am. Chem. Soc.*, *135*, 7462–7473, 2013.)

3.2.2 *Friedel–Crafts alkylation of indoles with α,β-unsaturated ketones*

Cinchona-based primary amines were first applied as chiral catalysts for the iminium ion activation of α,β-unsaturated ketones in 2007.[36,37] Thus, alkylation product **6** was obtained with moderate optical yield in the reaction between indole **4** and ketone **5** catalyzed by cinchona based amine **3** in the presence of double amount of trifluoroacetic acid (Scheme 3.12).[37]

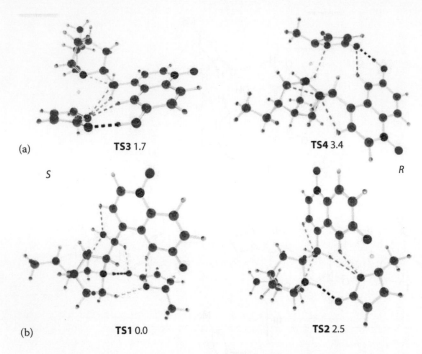

Figure 3.8 Optimized structures (M05-2x/6-31+G(d)/SMD) and relative energies (kcal/mol) of the protonation stage transition states in Scenarios **I** (a) and **III** (b). The weak noncovalent interactions important for the stereoselection are shown by orange dotted lines. (Adapted with permission from Xue, X. S. et al., *J. Am. Chem. Soc.*, 135, 7462–7473. Copyright 2013 American Chemical Society.)

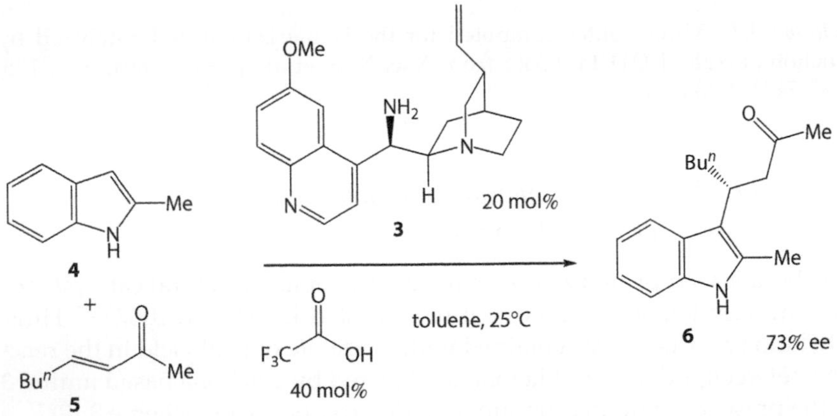

Scheme 3.12 Alkylation of indole **4** catalyzed by cinchona-based amine **3**.

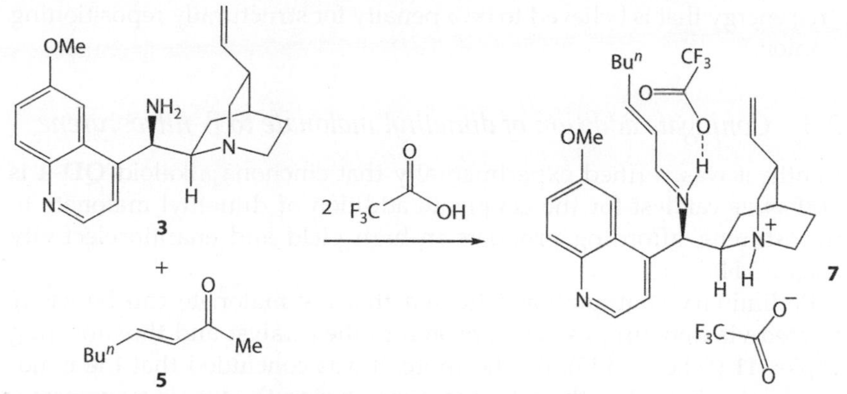

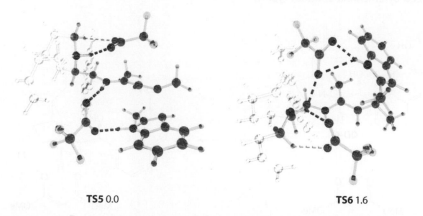

Scheme 3.13 Initial formation of iminium cation 7.

According to a "generally accepted mechanism,"[38] the reaction was considered to proceed via the initial formation of iminium ion **7** stabilized by two trifluoroacetate anions (Scheme 3.13).[39]

Experimental studies showed that two equivalents of the acid for one equivalent of catalysis are essential to obtain highest optical yields. It was therefore concluded that coordination of the counterion is an important stereoregulating factor. Thus, one of the coordinated trifluoroacetate anions in the **TS5** (Figure 3.9) is keeping two substrates together, whereas the second one blocks effectively one of the prochiral planes of the ketone. The **TS6** leading to the opposite enantiomer was computed to be 1.6 kcal/mol higher

TS5 0.0   **TS6** 1.6

Figure 3.9 Optimized transition-state structures for the addition of indole **4** to the iminium ion assembly **7**. Relative free energies ($\Delta\Delta G^{318}$) given in kcal/mol; M06-2X/6-311++G(d,p)/SMD//M06-2X/6-31G(d). (Adapted with permission from Moran, A. et al., *J. Am. Chem. Soc.*, 135, 9091–9098. Copyright 2013 American Chemical Society.)

in free energy that is believed to be a penalty for structurally repositioning the anion.[39]

3.2.3   Conjugate addition of dimethyl malonate to β-nitrostyrene

Recently, it was verified experimentally that cinchona alkaloid **QD-4** is an effective catalyst for the conjugate addition of dimethyl malonate to β-nitrostyrene, affording products in high yield and enantioselectivity (Scheme 3.14).[33]

Preliminary computations showed that the malonate can be easily activated via donating its acidic proton to the catalyst and thus forming complex **11** (Scheme 3.15). Furthermore, it was concluded that the malonate remains bound to the nitrogen atom during the whole transformation and that **9** cannot be activated solely by the alkoxy hydroxyl group due to its insufficient acidity.[40]

Hence, six pathways were analyzed in detail involving activation of the nitrostyrene either by single phenoxy hydroxyl or by both hydroxyl

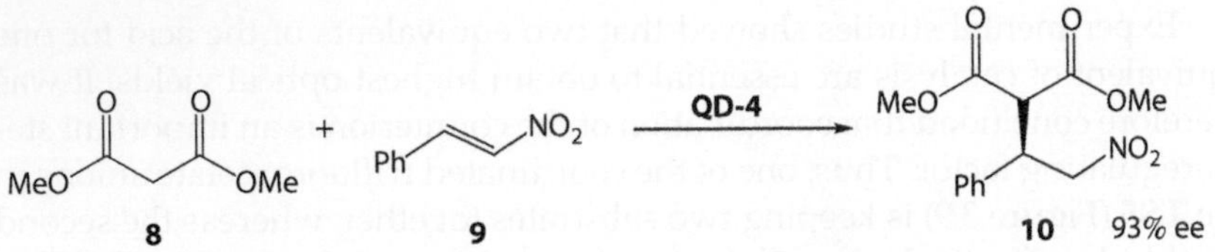

Scheme 3.14 Conjugate addition of dimethyl malonate to β-nitrostyrene catalyzed by cinchona alkaloid **QD-4**.

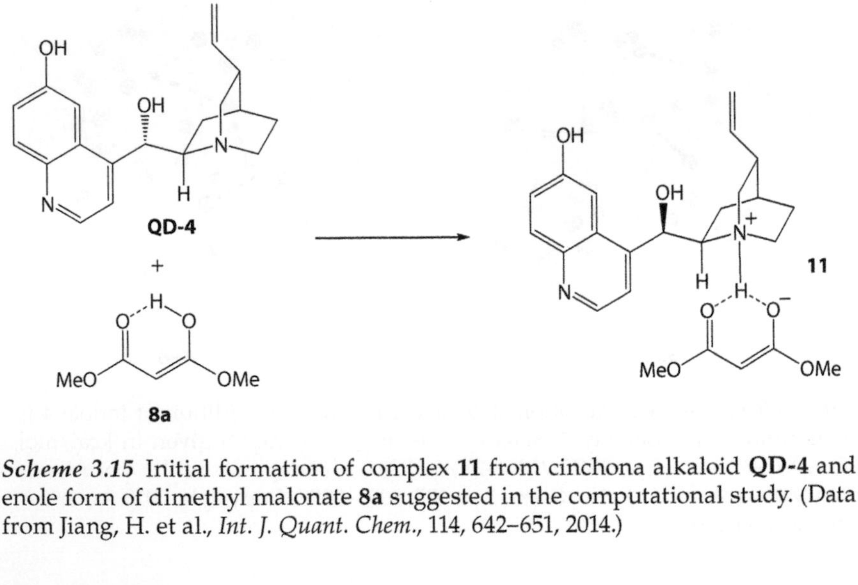

Scheme 3.15 Initial formation of complex **11** from cinchona alkaloid **QD-4** and enole form of dimethyl malonate **8a** suggested in the computational study. (Data from Jiang, H. et al., *Int. J. Quant. Chem.*, 114, 642–651, 2014.)

groups simultaneously. As a result, two lowest energy pathways leading to the opposite enantiomers were located, both involving only phenoxy hydroxyl of the catalyst for the substrate activation. The secondary protonation step was found to be a rate- and stereo-determining one.[40] Structural reasons for the stereoselection were not discussed.

3.2.4   Fluorination of cyclic ketones

The quinidine-derived primary amine **12** was identified as a highly selective catalyst for the α-monofluorination of cyclic ketones, for example, **14** with *N*-fluorobenzenesulfonimide **15** (Scheme 3.16).[41]

Computations were performed for the *N*-fluoroquinuclidinium ion **17**[42]; formation of such species is known to facilitate fluorine transfer significantly.[43] It is noted that the conformations of the seven-membered ring transition states **TS7** and **TS8** (Figure 3.10) are different and comparable to the known conformations of cycloheptane: in the **TS7** the ring adopts a chair conformation, whereas in the **TS8** the boat conformation is taken to expose the opposite enantioface.

The authors suggest that this conformational difference itself is the reason making **TS7** (6.8 kcal/mol) more stable than **TS8**.[42] From our point of view, a more reasonable explanation is the more effective stabilization of the transferred fluorine atom by noncovalent C–H...F interactions in the case of the **TS7**. In this transition state the fluorine atom has seven close contacts with the hydrogens, the shortest one (2.28 Å) being with the axial proton on C9 adjacent to the quinoline substituent.

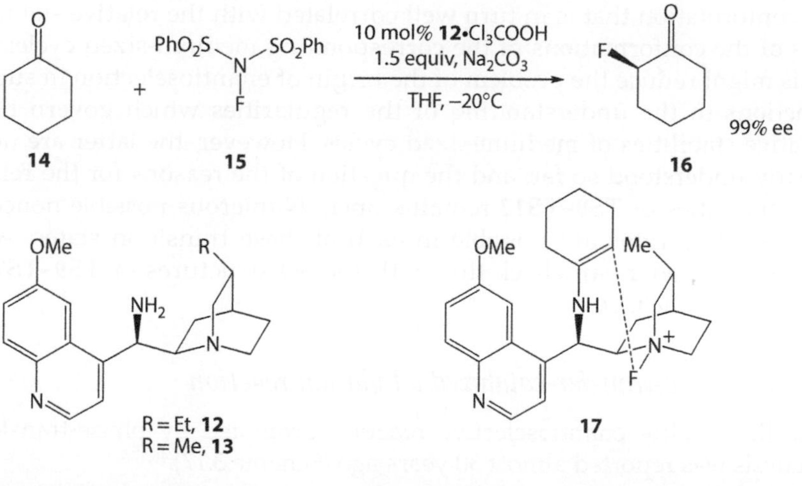

Scheme 3.16 Enantioselective monofluorination of cyclic ketones. (Data from Kwiatkowski, P. et al., *J. Am. Chem. Soc.*, 133, 1738–1741, 2011; Lam, Y.-h. and Houk, K. N., *J. Am. Chem. Soc.*, 136, 9556–9559, 2014.)

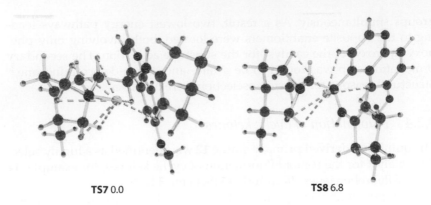

TS7 0.0   **TS8** 6.8

Figure 3.10 Optimized structures of **TS7** and **TS8** (B3LYP-D3(BJ)/def2-TZVPP-IEFPCM (THF)//B3LYP/6-31G(d)-IEF-PCM (THF)). (Data from Lam, Y.-H. and Houk, K. N., *J. Am. Chem. Soc.*, 136, 9556–9559, 2014.)

On the other hand, in the **TS8** the fluorine atom must move away from this proton, and the whole number of the stabilizing noncovalent C–H…F interactions is reduced to five. One of these interactions is with the *ortho*-proton of the quinolone substituent that explains its axial orientation.

The same authors studied recently the origins of enantioselection in intramolecular aldol reactions catalyzed by cinchona alkaloids.[44] Computations revealed the competing transition states (**TS9-TS12**), the relative energies of which correlated well with the observed handedness and level of enantioselection. Similarly to the previous study, the authors noticed the correlation of the stability of a transition state with its conformation that is in turn well correlated with the relative stabilities of the conformations of the corresponding medium-sized cycles.[45] This might reduce the problem of the origin of enantioselection in such reactions to the understanding of the regularities which govern the relative stabilities of medium-sized cycles. However, the latter are not clearly understood so far, and the question of the reasons for the relative stabilities of **TS9–TS12** remains open. Numerous possible noncovalent interactions are possible in each of these transition states. An interested reader can check this with the 3D structures of **TS9–TS12** available in the CD.

3.2.5 Phase-transfer-catalyzed alkylation reaction

The first highly enantioselective reaction promoted by phase-transfer catalysis was reported almost 30 years ago (Scheme 3.17).[46,47]

Early mechanistic suggestions on the origin of enantioselectivity suggested involvement of π–π stacking[47] or the interaction of the cinchonidinium cation with the anion of a reagent.[48]

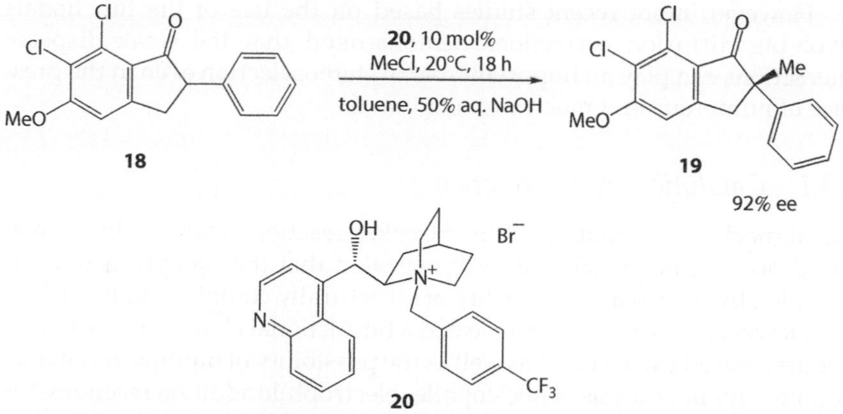

Scheme 3.17 Enantioselective phase-transfer alkylation catalyzed by **20**.

In a recent computational analysis of the reaction shown in Scheme 3.17, it was suggested that substrate **18** is deprotonated by the base present in the system and the resulting indanone anion is forming a stable ion pair with the cinchonium cation which was supposed to keep its hydroxyl group untouched.[49]

In neither of 12 located ion pair structures, a π–π stacking interaction was found. The transition states for the enantioselective alkylations were characterized only for the four most stable ion pairs. Only two of these transition states, **TS13** and **TS14**, belonged to the reactions with activation barriers lower than 30 kcal/mol.

Linear dependence of the relative free energy of a transition state versus the distance between the anion Cl⁻ and the positively charged nitrogen atom of the cinchonium cation was interpreted as an argument in favor of the electrostatic stabilization of the transition states. The most stable **TS13** leading to the experimentally observed *S*-enantiomer had the lowest chloride–nitrogen distance. However, the structural reasons allowing the maximal approach of the chloride anion to the nitrogen atom were not analyzed.

3.3   Urea and thiourea-based catalysts

Thioureas (and less often ureas) as organocatalysts were introduced into synthetic practice by Jacobsen[50,51] and Schreiner.[52–54] Further development included the design of bifunctional catalysts combining two thiourea subunits or a thiourea group and chiral amines.[55–57]

Especially these more effective bifunctional catalysts are capable of formation of numerous strong hydrogen bonds with the substrates of the catalytic reactions. In earlier studies, these strong hydrogen bonds were considered to be a main stereo-regulating factor in the corresponding catalytic reactions.

However, more recent studies based on the use of the functionals involving diffusion corrections demonstrated that the weak disperse interactions can play an important role in stereoselection even in the presence of numerous and much stronger bonds.

3.3.1 Catalytic Strecker reaction

Initial mechanistic analysis of the Strecker reaction catalyzed by a urea-based organocatalyst (Scheme 3.18) revealed that the catalytic activity is provided by the urea functionality of structurally complex catalyst **1**.[51]

However, further studies revealed a bifunctional character of urea and thiourea-based catalysts[58–60] as well as the possibility of multiple mechanistic pathways in catalysis of nucleophile–electrophile addition reactions.[61–63]

Simplified but sufficiently effective (thio)urea catalysts **4a** and **4b** were used in the hydrocyanation reaction (Scheme 3.19) that was subjected to a combined experimental and computational study.[60]

Hammett correlation of the reactivity of the substrates bearing substituents with various electronic properties allowed to conclude that the mechanism involves rate-limiting formation of an iminium ion intermediate **5** (Scheme 3.20).

Computational analysis of the catalytic cycle was made for the simplified catalyst **6**. The possibilities of either HCN or HNC addition have been analyzed separately; however, the key transition states were found to be similar in structures and in relative energies for both mechanisms.

For the analysis of the origin of enantioselectivity computations were made for a more realistic model catalyst **7**.

6   **7**

The rearrangement of the ion pair **8** yielding **9** was identified as a stereo-regulating step. The next step affording the catalyst–product **10** was computed to be also stereoselective, and leading to the same enantiomer (Scheme 3.21).

It was concluded that enantioselectivity is controlled by different degrees of iminium ion stabilization via H-bonding interactions between the iminium ion NH proton and both the amide carbonyl group and the thiourea-bound cyanide ion.[60] However, distinct structural features responsible for that more effective stabilization are not easy to specify. The 3D structures of **TS1R**, **TS2R**, **TS1S**, and **TS2S** can be found in the CD.

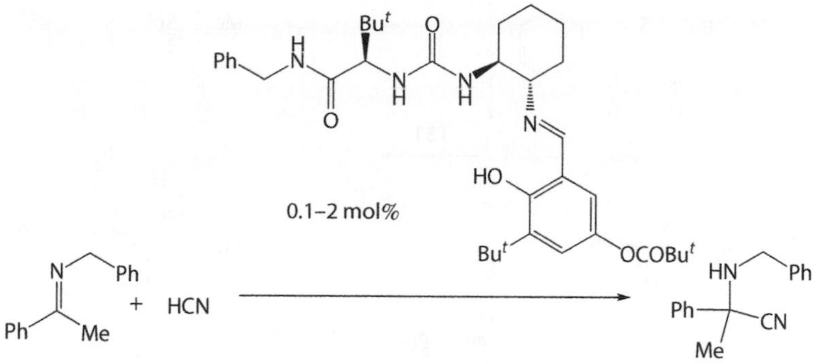

Scheme 3.18 Strecker reaction catalyzed by urea-based catalyst **1**.

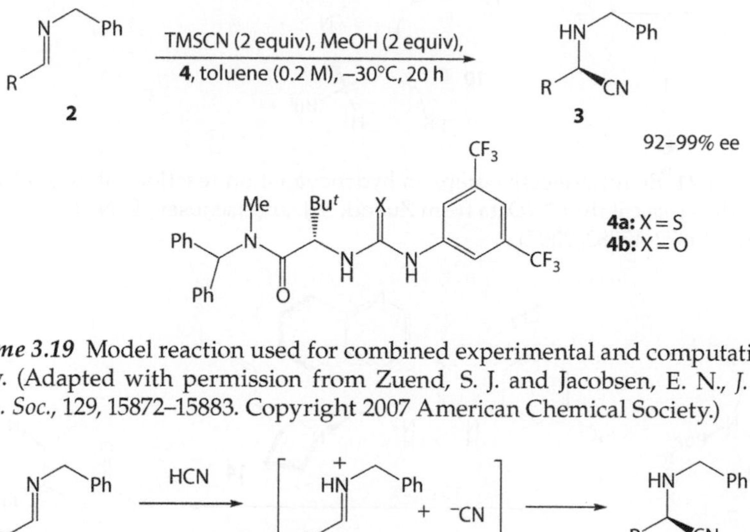

Scheme 3.19 Model reaction used for combined experimental and computational study. (Adapted with permission from Zuend, S. J. and Jacobsen, E. N., *J. Am. Chem. Soc.*, 129, 15872–15883. Copyright 2007 American Chemical Society.)

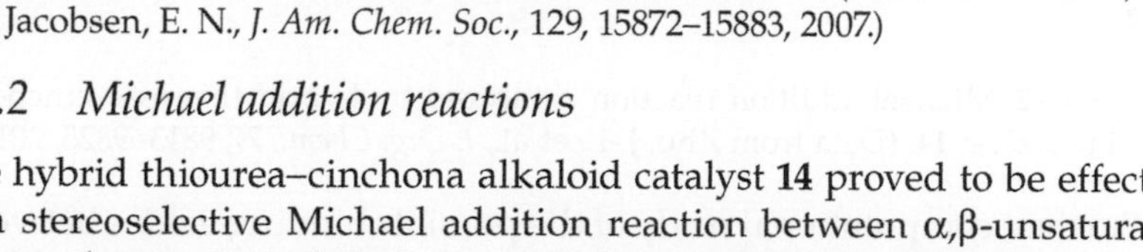

Scheme 3.20 Formation of iminium ion intermediate **5**. (Data from Zuend, S. J. and Jacobsen, E. N., *J. Am. Chem. Soc.*, 129, 15872–15883, 2007.)

3.3.2   Michael addition reactions

The hybrid thiourea–cinchona alkaloid catalyst **14** proved to be effective in a stereoselective Michael addition reaction between α,β-unsaturated γ-butyrolactam **11** and chalcone **12** (Scheme 3.22).[64] The following mechanistic study[65] addressed the issue of the origin of stereoselection.

Conformational analysis of the catalyst molecule showed that a conformation with all three nitrogen atoms facing the same direction (shown in the Scheme 3.20) is a stable minimum, and hence bifunctional simultaneous activation of both reagents is possible.

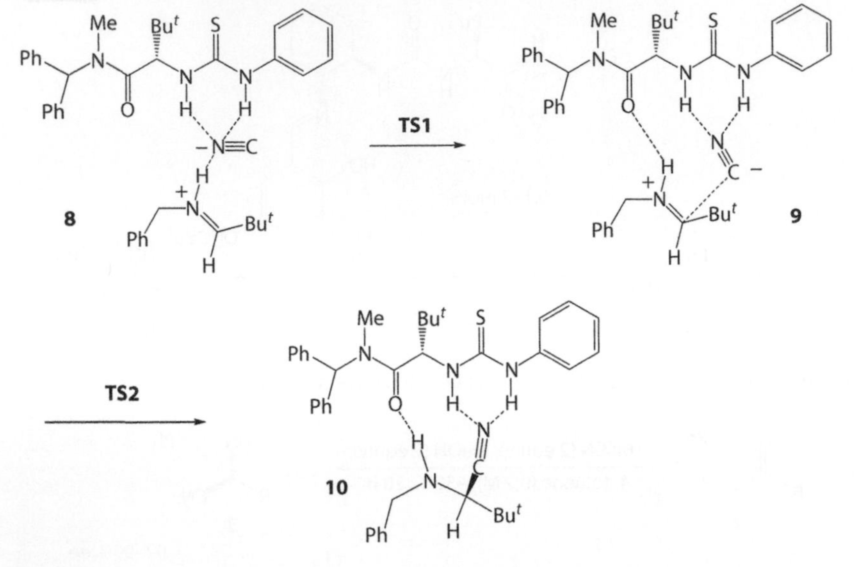

Scheme 3.21 Enantioselective steps in hydrocyanation reaction catalyzed by the model thiourea catalyst **7**. (Data from Zuend, S. J. and Jacobsen, E. N., *J. Am. Chem. Soc.*, 129, 15872–15883, 2007.)

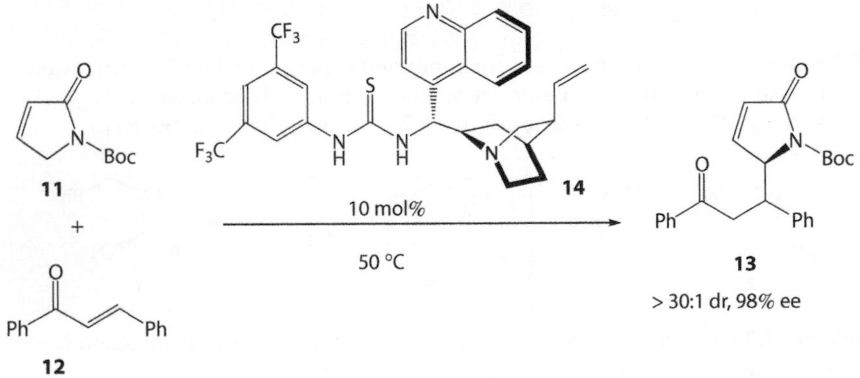

Scheme 3.22 Michael addition reaction catalyzed by hybrid thiourea–cinchona alkaloid catalyst **14**. (Data from Zhu, J.-L. et al., *J. Org. Chem.*, 77, 9813–9825, 2012.)

Further, computational study of the possible binding modes between **14** and enolic form of **11** revealed the most facile route for the proton transfer (Scheme 3.23). Initial formation of the adduct **16** is followed by a barrierless proton transfer resulting in the ion pair **16** in which all three NH protons are involved in multiple intramolecular hydrogen bonds.

The chalcone **12** is supposed to coordinate with one of the thiourea NH protons to yield trimolecular complex **17** (Scheme 3.24). The rate-limiting

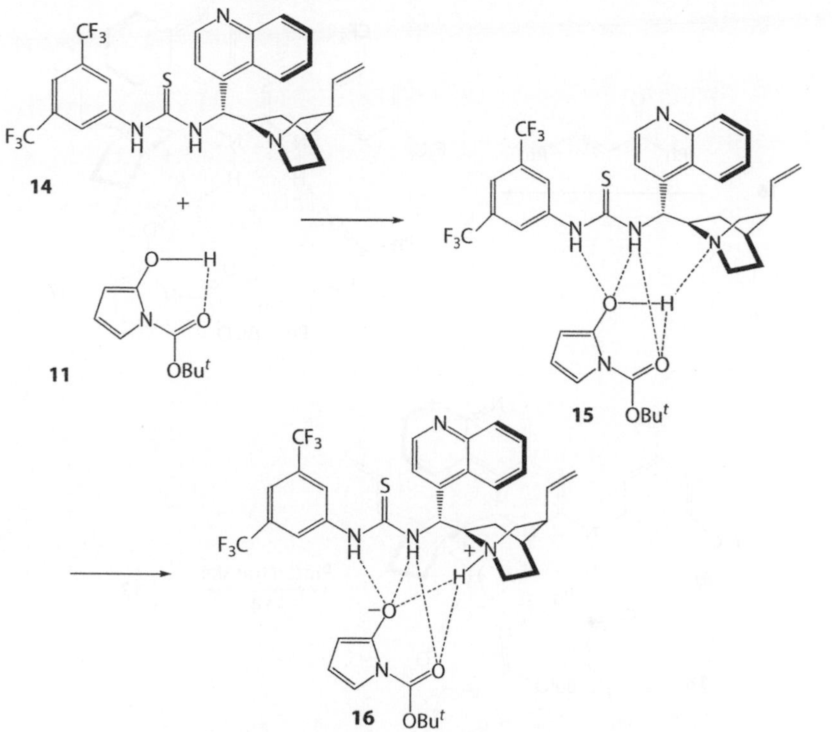

Scheme 3.23 Formation of the catalyst–substrate complex **16**. (Data from Zhu, J.-L. et al., *J. Org. Chem.*, 77, 9813–9825, 2012.)

and stereo-determining C–C bond formation leads to the ion pair **18**, and the following barrierless proton transfer yields the reaction product **13**.

Stereoselectivity analysis involved optimization of four transition states for the C–C bond formation step leading to different diastereomers (Figure 3.11). Computed selectivity (100% ee and 60:1 dr) correlated well with the experimental values (98% ee and >30:1 dr).

Structures of the corresponding transition states **TS3** (*RS* 6.5 kcal/mol), **TS4** (*SR* 10.4 kcal/mol), **TS5** (*SS* 9.1 kcal/mol), and **TS6** (*RR* 15.2 kcal/mol) can be found in the CD. However, one should keep in mind that the dispersion was not accounted for in these computations, whereas the structures of the transition states imply numerous possibilities of weak intramolecular interactions.

3.3.3   *Enantioselective decarboxylative protonation*

The same catalyst **14** was found to promote enantioselective decarboxylative protonation reaction.[66,67] Recent computational study used DFT/PM3

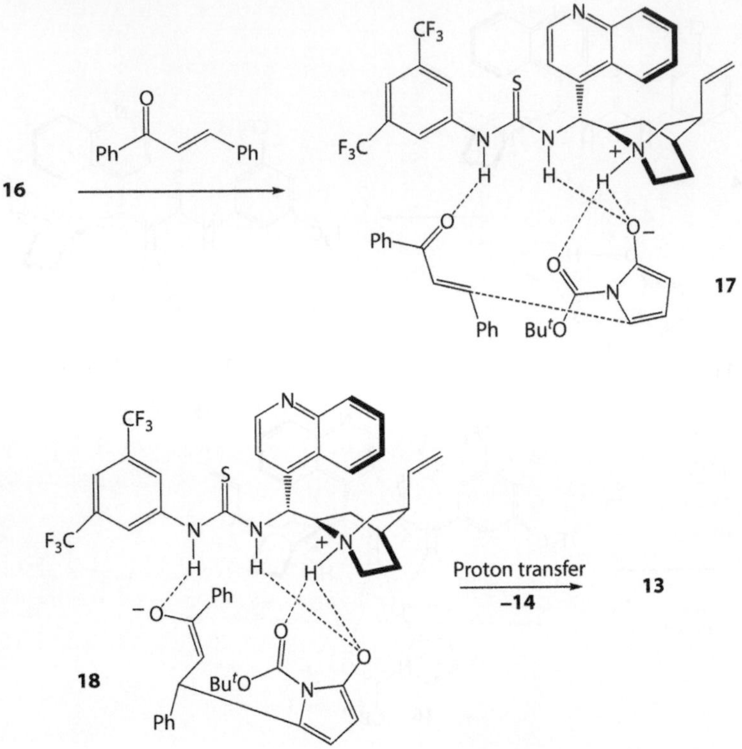

Scheme 3.24 Stereoselective C–C bond formation step.

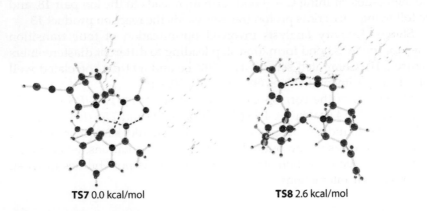

TS7 0.0 kcal/mol   **TS8** 2.6 kcal/mol

Figure 3.11 Diastereomeric transition states for the direct protonation of the enolate carbon. (Adapted with permission from Sengupta, A. and Sunoj, R. B., *J. Org. Chem.*, 77, 10525–10536. Copyright 2012 American Chemical Society.)

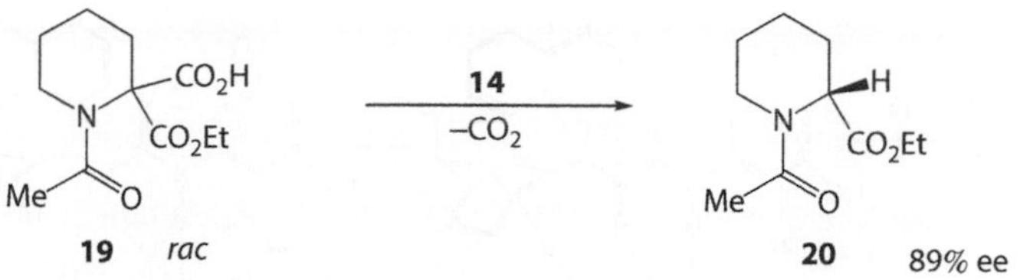

Scheme 3.25 Hybrid thiourea–cinchona-catalyzed asymmetric decarboxylative protonation. (Data from Blanchet, J. et al., *Eur. J. Org. Chem.*, 5493, 2008.)

ONIOM partition scheme to analyze mechanism and origin of enantioselectivity in this reaction (Scheme 3.25).[68]

Conformational studies revealed numerous minima in the 15 kcal/mol range. Mechanistic studies were performed with the conformer with *syn*-orientation of the N–H groups (relative free energy with respect to the conformational minimum 3.06 kcal/mol), since the simultaneous availability of two hydrogen bond donor N–H groups implies more effective substrate binding.

Five modes of binding between **19** and *syn*-N–H conformer of **14** were identified with the relative energies of the most and least stable isomer being 0.0 and 8.9 kcal/mol, respectively. The most stable complex **21** was used for the study of the reaction mechanism (Scheme 3.26).

Release of CO_2 from **21** yields enolate **22**, proton transfer in the latter gives enol **23**. Simple tautomerization that could easily afford the reaction product was computed to be unreasonably high in energy; hence, the enol **23** was computed to rearrange in **24** via reorganization of the hydrogen bond networks accompanied by the proton returning to the quinuclidine unit, thus forming a carbanionic center. The subsequent proton transfer gives the catalyst–product complex **25** (Scheme 3.26).

The computed mechanism (Scheme 3.26) implied that the direct protonation of the enolate carbon in **24** is a stereo-determining step of the reaction, since it could occur through either of its prochiral faces.

Transition states for the enantioselective intramolecular protonation and their relative stabilities are shown in Figure 3.11. The relative stability of the **TS7** is supposed to be stipulated by more effective hydrogen binding. However, the possibility of numerous weak interactions (see Figure 3.11) might be not taken into account during optimizations using ONIOM DFT/PM3 approach.

3.3.4 Henry reaction

Recently, a Henry reaction between aldehydes (e.g., benzaldehyde **26**) and nitroalkanes (e.g., nitromethane **27**) catalyzed by cyclophane-based thiourea **28** was reported (Scheme 3.27).[69]

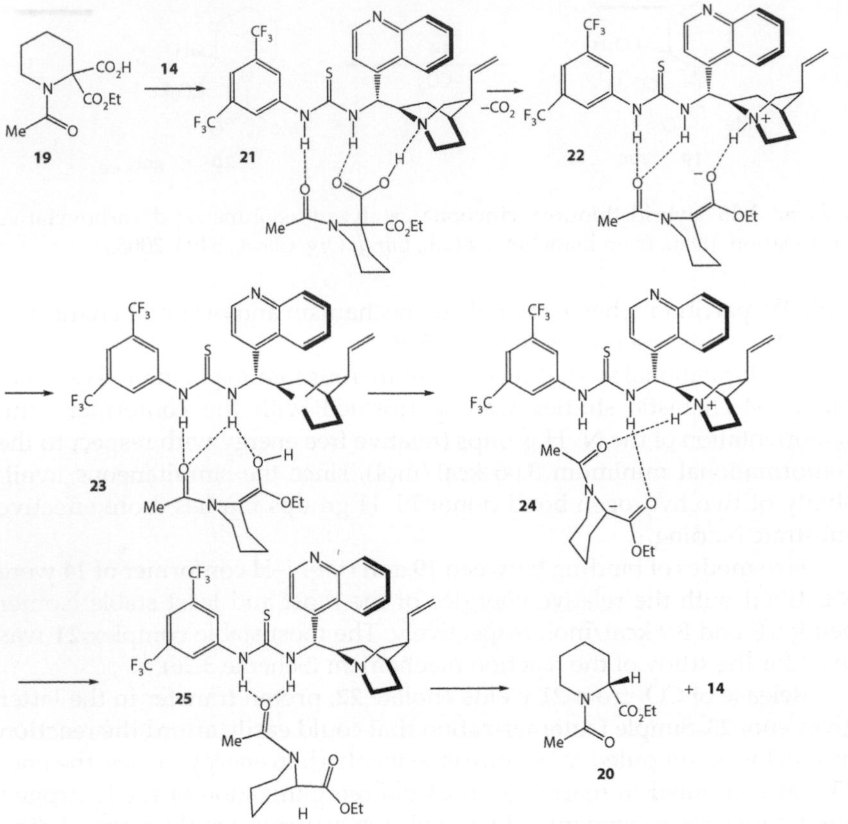

Scheme 3.26 Computed relatively facile mechanism of the enantioselective decarboxylative protonation. (Adapted with permission from Sengupta, A. and Sunoj, R. B., *J. Org. Chem.*, 77, 10525–10536. Copyright 2012 American Chemical Society.)

In a computational study catalyst **30** bearing two methyls instead of two bis-(trifluoromethyl)phenyl groups was employed[70] (Scheme 3.27).

In the computational study, a possibility of simultaneous binding of both reagents with hydrogen bonds with both N–H protons in different thiourea units was investigated. It was found, however, that the most stable complex of this kind is significantly (by 6 kcal/mol) less stable than a complex of nitronate anion with the catalyst and free benzaldehyde, whereas the binding of free benzaldehyde is very weak. It was concluded therefore that the complex in which both reagents are bound to the catalyst is not an important intermediate due to its low concentration and aldehyde **26** preferentially remains unbound before the formation of the transition state.[70]

Further search of the corresponding transition states afforded **TS9** and **TS10** with the relative stabilities correctly predicting the experimentally

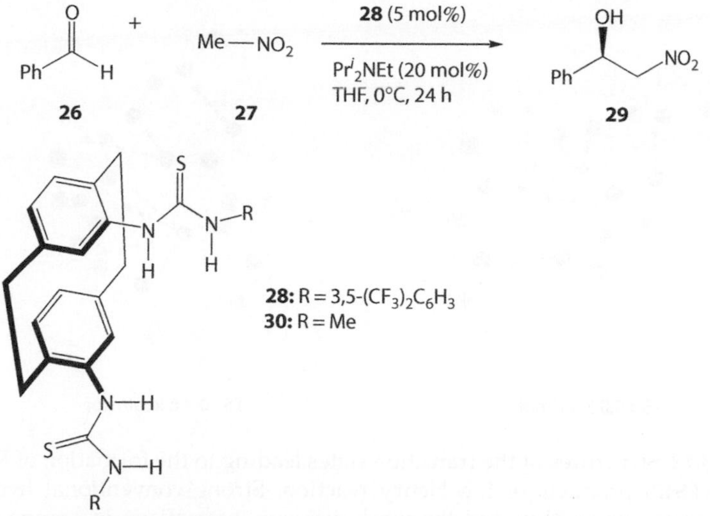

Scheme 3.27 Cyclophane-based bisthiourea-catalyzed asymmetric Henry reaction. (Data from Kitagaki, S. et al., *Chem. Commun.*, 49, 4030–4032, 2013; Breugst, M. and Houk, K. N., *J. Org. Chem.*, 79, 6302–6309, 2014.)

observed handedness of the reaction (Figure 3.12), albeit the level of enantioselection is strongly overestimated (presumably due to the reversibility of the reaction).[70]

The reasons for a significant difference in stability between **TS9** and **TS10** are unclear. The authors carefully checked any possible computational errors by reoptimizing these structures in numerous basis sets and with various ways of description of the dispersion effects to get almost identical difference in stabilities.

Either **TS9** or **TS10** contains four strong conventional hydrogen bonds and three C–H...O or C–H...H–C interactions. Apparently, more careful analysis might provide valuable information on the relative strength of various intramolecular interactions. The failure to recognize structural factors making one of the diastereomeric transition states 3.6 kcal/mol more stable than another one, displays limits in our capability to elucidate the origins of enantioselection and apply them for a "rational catalysis design."

3.3.5 α-Hydroxylation of β-ketoesters

Recently, an α-hydroxylation reaction of tetralone-derived ketoesters **31** with cumene hydroperoxide **32** in the presence of bis-urea guanidine based catalyst **33** was developed (Scheme 3.28).[71]

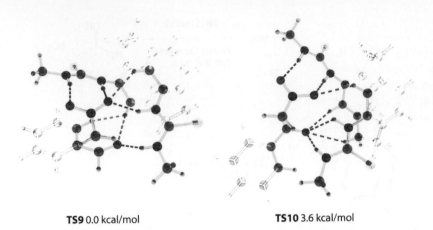

TS9 0.0 kcal/mol **TS10** 3.6 kcal/mol

Figure 3.12 Structures of the transition states leading to the formation of *R* (**TS9**) and *S* (**TS10**) products of the Henry reaction. Strong conventional hydrogen bonds are shown in blue, and the weak disperse interactions in orange dotted lines. (Adapted with permission from Breugst, M. and Houk, K. N., *J. Org. Chem.*, 79, 6302–6309. Copyright 2014 American Chemical Society.)

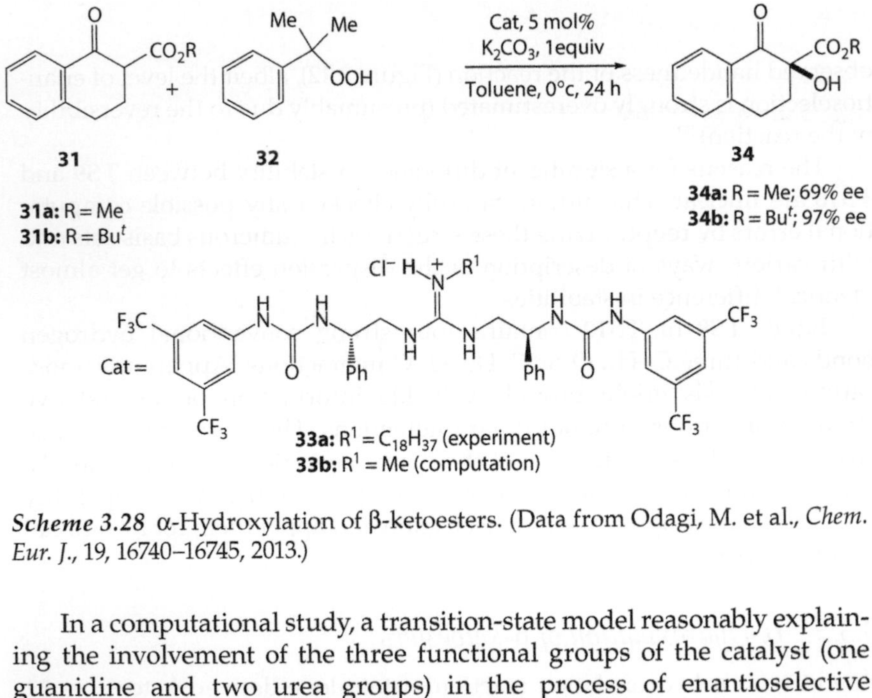

O
CO_2R
+
Me
Me
OOH
Cat, 5 mol%
K_2CO_3, 1equiv
Toluene, 0°c, 24 h
CO_2R
OH
31
32
34
34a: R = Me; 69% ee
34b: R = Bu^t; 97% ee
31a: R = Me
31b: R = Bu^t
Cat =
Cl^- H + R^1
F_3C
CF_3
Ph
Ph
33a: $R^1 = C_{18}H_{37}$ (experiment)
33b: R^1 = Me (computation)

Scheme 3.28 α-Hydroxylation of β-ketoesters. (Data from Odagi, M. et al., *Chem. Eur. J.*, 19, 16740–16745, 2013.)

In a computational study, a transition-state model reasonably explaining the involvement of the three functional groups of the catalyst (one guanidine and two urea groups) in the process of enantioselective α-hydroxylation was developed.[72] In this model, the orientation of the ketoester enolate is controlled by hydrogen bonding of its two carbonyl groups with four N–H units in guanidinium and urea groups of the catalyst. The second urea unit binds **32** via the oxygen-attached carbon. Numerous disperse interactions were found to stabilize both of the transition states with a larger number of those found in the **TS11** preceding the experimentally observed S-enantiomer of **34**. In addition, the **TS12** leading to the wrong isomer was found to be destabilized by its ester group that is embedded in the narrow space constructed by the Ar groups of the catalyst. This explains the effect of the R substituent on the level of enantioselection (Scheme 3.28) which was reasonably reproduced in the computations.[72] The structures of **TS11** and **TS12** are too cumbersome for a meaningful illustration; their 3D models can be found in the CD.

Of interest is the significant difference reported for the structures of the **TS11** optimized by two different functionals with and without accounting for dispersion (Figure 3.13). The C–H...F interaction showed

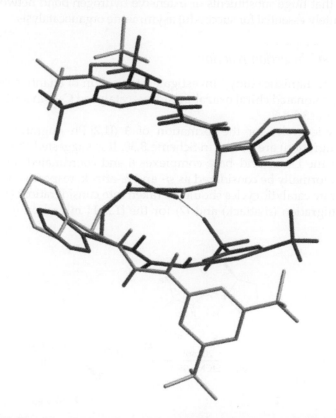

Figure 3.13 The 3D superposition of the catalyst structures in **TS11** (light tube: B3LYP/6-31G*, dark tube: M05-2X/6-31G*). (Adapted with permission from Odagi, M. et al., *J. Am. Chem. Soc.*, 137, 1909–1915. Copyright 2015 American Chemical Society.)

by a dotted line on Figure 3.13 remains completely unseen in the B3LYP/6-31G*-optimized structure that results in completely different conformation of a large part of the catalyst molecule. This result illustrates the importance of caution necessary in the interpretation of the computations carried out without accounting for the dispersion effects.

3.4   N-Protonated chiral oxazaborolidine

Chiral oxazaborolidine catalysts were applied in various enantioselective transformations including reduction of highly functionalized ketones,[73,74] oximes[75] or imines,[76] Diels–Alder reactions,[77,78] cycloadditions,[79] Michael additions,[80] and other reactions. These catalysts are surprisingly small molecules compared to the practically efficient chiral phosphoric acids, cinchona alkaloids, or (thio)ureas; hence, their effectiveness in asymmetric catalysis demonstrates that huge substituents or extensive hydrogen bond networks are not absolutely essential for successful asymmetric organocatalysis.

3.4.1   C–C insertion reaction

A recent mechanistic study[81] investigated the origin of enantioselectivity in the N-protonated chiral oxazaborolidine-catalyzed C–C insertion reaction (Scheme 3.29).[82]

Catalytic cycles for the formation of **3** ([1,2]-Ph migration) and **4** ([1,2]-H migration) are shown in Scheme 3.30. It is suggested that the formation of the Lewis acid–base complexes **6** and coordinated intermediates **7** can formally be considered as *si*- and *re*-attack, respectively. Hence, two separate catalytic cycles should be taken into consideration: (1) for the [1,2]-Ph migration (*si*-attack) and (2) for the [1,2]-H migration (*re*-attack).

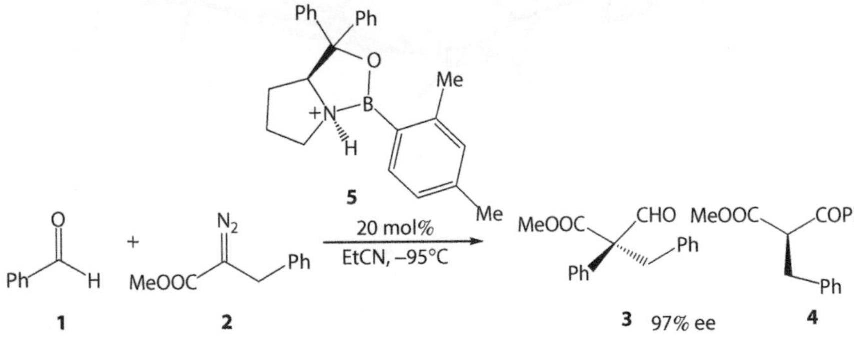

Scheme 3.29 N-Protonated chiral oxazaborolidine-catalyzed C–C (C–H) insertion reaction. (Data from Gao, L. et al., *J. Am. Chem. Soc.*, 135, 14556–14559, 2013.)

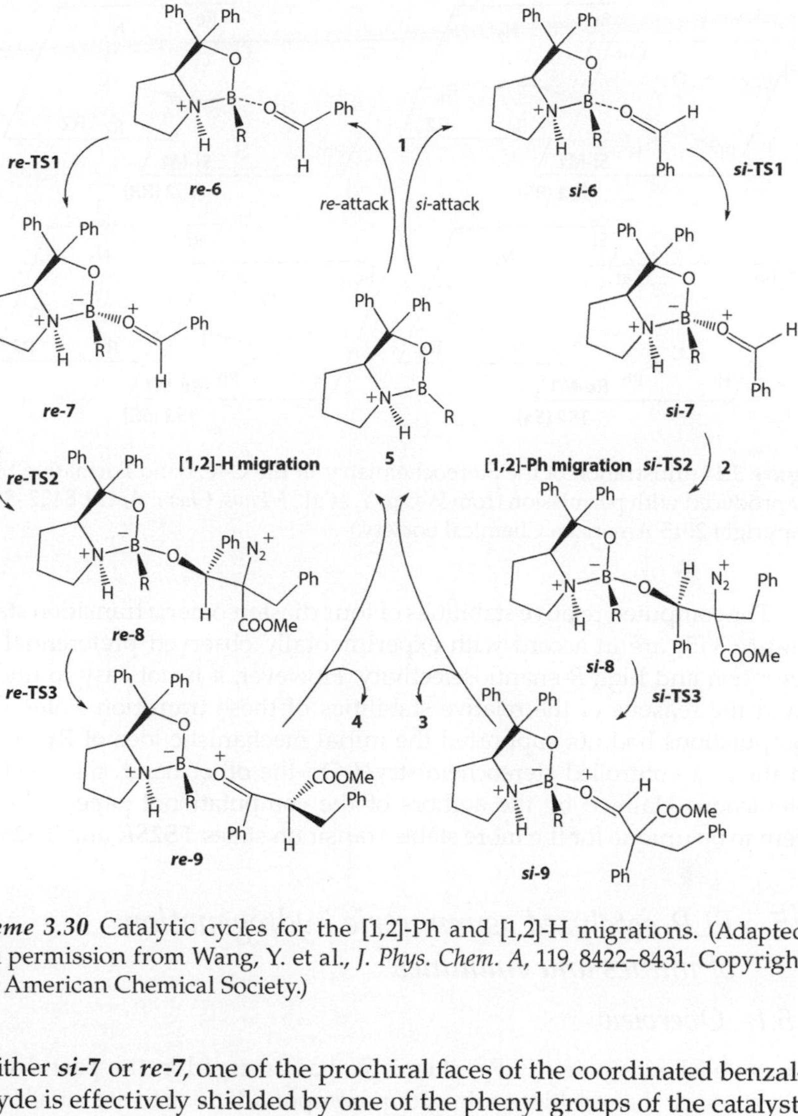

Scheme 3.30 Catalytic cycles for the [1,2]-Ph and [1,2]-H migrations. (Adapted with permission from Wang, Y. et al., *J. Phys. Chem. A*, 119, 8422–8431. Copyright 2015 American Chemical Society.)

In either *si-7* or *re-7*, one of the prochiral faces of the coordinated benzaldehyde is effectively shielded by one of the phenyl groups of the catalyst. This makes possible approach of the diazoester **2** only from another side, which is not shielded by the phenyl group.

The migration of the phenyl group or the proton was computed to occur in a concerted but asynchronous process. The concreteness of this step implies serious limitations on the relative orientation of the two reactants leaving in each case only two stereochemical possibilities for the formation of the C–C bond between coordinated benzaldehyde and diazoester **2** (Figure 3.14). Hence, totally four reaction pathways were considered in the computational study.

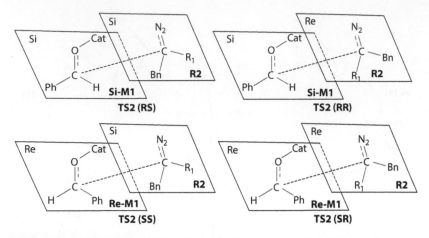

Figure 3.14 Illustration of the Stereochemistry in the C–C Bond Formation Step. (Reproduced with permission from Wang, Y. et al., *J. Phys. Chem. A*, 119, 8422–8431. Copyright 2015 American Chemical Society.)

The computed relative stabilities of four diastereomeric transition states (Figure 3.15) are in accord with experimentally observed preferential Ph migration and high *R*-enantioselectivity. However, it is not easy to understand the reasons of the relative stabilities of these transition states. The computations had not supported the initial mechanistic idea of Ryu et al. on the π–π controlled stereochemistry.[82] On the other hand, the C–H...π interactions claimed by the authors of the computational paper[81] do not seem to be unique for the more stable transition states **TS2SR** and **TS2SS**.

3.5   FLP-catalyzed asymmetric hydrogenation of imines and enamines

3.5.1   Overview

Frustrated Lewis pairs (FLPs) chemistry has emerged in the past decade as a strategy that enables main group compounds to activate small molecules.[83] The idea of FLP is based on the combinations of Lewis acids and bases that are sterically prevented from forming classical Lewis acid–base adducts. On the other hand, these FLPs have sufficient Lewis acidity and basicity available for interaction with a third molecule. Starting from stoichiometric reactions, the concept has been applied to catalytic hydrogenation of olefins, allene–esters, enones, imines, enamines, *N*-heterocycles, carbonyl compounds, tandem hydroamination/hydrogenation of terminal alkynes, and others.[83–86] Typical catalyst loadings are 5%–20%, but some reactions are carried out under mild reaction conditions (rt, small hydrogen

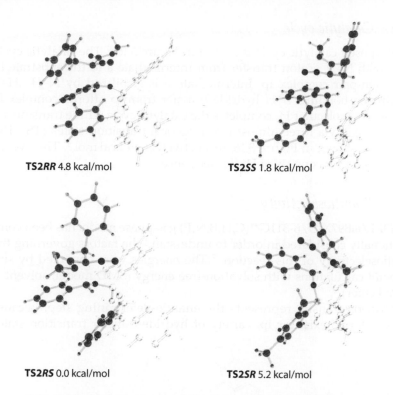

TS2RR 4.8 kcal/mol
TS2SS 1.8 kcal/mol
TS2RS 0.0 kcal/mol
TS2SR 5.2 kcal/mol

Figure 3.15 Optimized structures of the diastereomeric transition states for the C–C bond formation accompanied by release of N_2. (Adapted with permission from Wang, Y. et al., *J. Phys. Chem. A*, 119, 8422–8431. Copyright 2015 American Chemical Society.)

(S)-catalyst:
+ H2
2 bar
(S)-catalyst
MTBE, 25°C
(R)
Conditions: 30 min, substrate (0.2 M)
S/C = 40
99% y., 99% ee

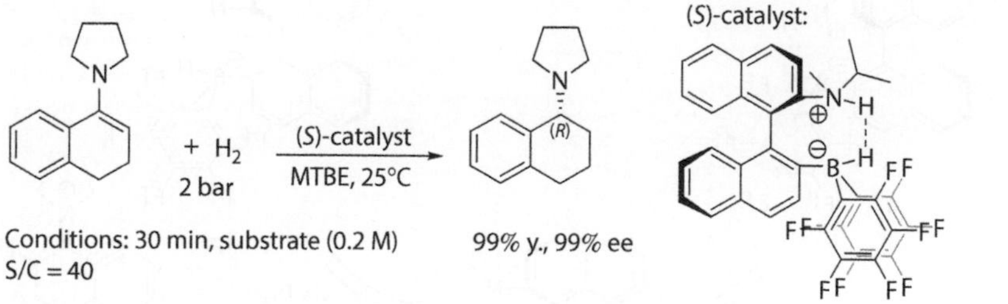

Scheme 3.31 FLP-catalyzed asymmetric hydrogenation of 1,2-dihydro -4-pyrrolidinonaphthalene.

pressure). Only recently, few asymmetric versions,[87–93] in particular for imine hydrogenations, were developed. Although the enantioselectivity is from moderate to good in vast majority of cases, few substrates (i.e., 1,2-dihydro-4-pyrrolidinonaphthalene, see Scheme 3.31) afforded chiral amines with 99% ee,[87] demonstrating further potential.

3.5.2 Catalytic cycle

The proposed catalytic cycle is shown in Figure 3.16.[87] The catalytic cycle starts with N–H proton transfer from intermediate **a** to the substrate to afford ion-pair complex **ip**. Intermediate **a** is stabilized by N–H...H–B dihydrogen bonding. B–H hydride transfer from **ip** affords complex **b**. Cleavage of molecular H_2 completes the catalytic cycle. Three transformations proceed via three almost isoenergetical transition states (TS$_1$, TS$_2$, and TS$_3$ as shown in Figure 3.16, respectively), ~14 kcal·mol^{-1}. The overall mechanism thus consists of H$^+$/H$^-$ sequence.

3.5.3 Enantioselectivity

Full DFT/ωB97X-D/6-311G**(C,H,B,N,F) gas-phase model has been computationally considered in order to understand the factors governing the enantioselectivity of this reaction.[87] The energies were corrected by single-point calculations with solvation-free energy (SMD model, solvent = diethyl ether).

Hydride transfer represents the enantio-determining step. Because isomeric forms exist for **ip**, variety of hydride-transfer transition states

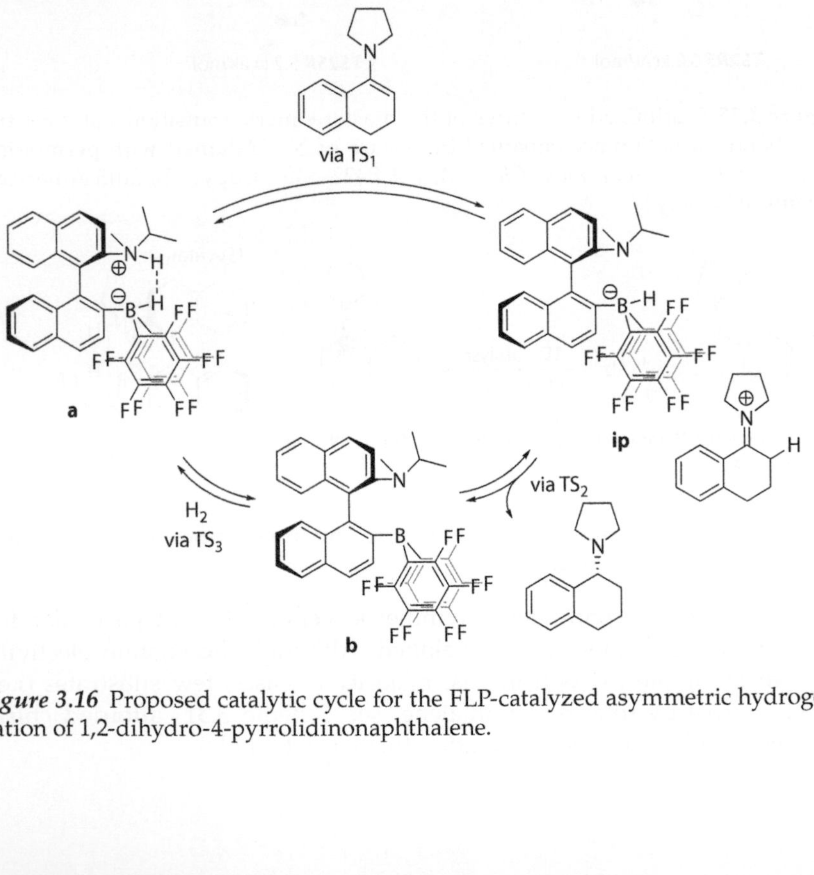

Figure 3.16 Proposed catalytic cycle for the FLP-catalyzed asymmetric hydrogenation of 1,2-dihydro-4-pyrrolidinonaphthalene.

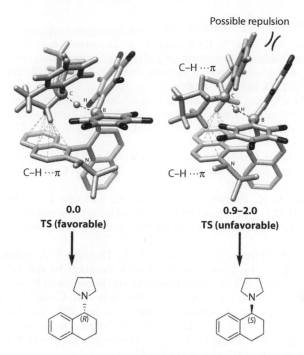

Figure 3.17 The most stable transition states leading to the (R) and (S) product enantiomers. Energy, kcal·mol⁻¹. Selected hydrogen atoms are omitted for clarity.

are possible. The most stable[87] transition states yielding the enantiomeric products are shown in Figure 3.17.

Both TSs are stabilized by series of weak C–H···π interactions between substrate and BINAP π–electrons. Slightly longer N–H and B–H distances in the unfavorable TS might be an indication of increased repulsive intermolecular forces and its higher energy, respectively. The origin of these forces could be repulsive interaction between aromatic ring of naphthalene and aromatic ring of one C_6F_5 group. Note that with respect to benzene the quadrupole moment is reversed for hexafluorobenzene.[94] The gas-phase optimized geometries predict a difference of 0.9–2.0 kcal/mol between the most stable transition states leading to the (R) and (S) product enantiomers depending on the calculation method. The sense of stereoselection is predicted correctly (preference for (R)-product), however, the computed enantiomeric excess (65%–93.8%) is lower than that observed experimentally for this reaction (99%). Thus, it is possible that solvent effect also contribute to the reaction enantioselectivity.

References

1. Roush, W. R.; Walts, A. E.; Hoong, L. K. Diastereo- and Enantioselective Aldehyde Addition Reactions of 2-allyl-1,3,2-Dioxaborolane-4,5-Dicarboxylic Esters: A Useful Class of Tartrate Ester Modified Allylboronates. *J. Am. Chem. Soc.* 1985, *107*, 8186–8190.
2. Brown, H. C.; Bhat, K. S.; Randad, R. S. Chiral Synthesis via Organoboranes. 21. Allyl- and Crotylboration of Alpha-Chiral Aldehydes with Diisopinocampheylboron as the Chiral Auxiliary. *J. Org. Chem.* 1989, *54*, 1570–1576.
3. Corey, E. J.; Yu, C.-M.; Lee, D.-H. A Practical and General Enantioselective Synthesis of Chiral Propa-1,2-Dienyl and Propargyl Carbinols. *J. Am. Chem. Soc.* 1990, *112*, 878–879.
4. Gonzalez, A. Z.; Roman, I. G.; Alicea, E.; Canales, E.; Soderquist, J. A. Borabicyclo[3.3.2] Decanes and the Stereoselective Asymmetric Synthesis of 1,3-Diol Stereotriads from 1,3-Diborylpropenes. *J. Am. Chem. Soc.* 2009, *131*, 1269–1273.
5. Althaus, M.; Mahmood, A.; Suarez, J. R.; Thomas, S. P.; Aggarwal, V. K. Application of the Lithiation–Borylation Reaction to the Preparation of Enantioenriched Allylic Boron Reagents and Subsequent In Situ Conversion into 1,2,4-Trisubstituted Homoallylic Alcohols with Complete Control over All Elements of Stereochemistry. *J. Am. Chem. Soc.* 2010, *132*, 4025–4028.
6. Kennedy, J. W. J.; Hall, D. G. Dramatic Rate Enhancement with Preservation of Stereospecificity in the First Metal-Catalyzed Additions of Allylboronates. *J. Am. Chem. Soc.* 2002, *124*, 11586–11587.
7. Lachance, H.; Xu, M.; Gravel, M.; Hall, D. G. Scandium-Catalyzed Allylboration of Aldehydes as a Practical Method for Highly Diastereo- and Enantioselective Construction of Homoallylic Alcohols. *J. Am. Chem. Soc.* 2003, *125*, 10160–10161.
8. Ishiyama, T.; Ahiko, T.; Miyaura, N. Acceleration Effect of Lewis Acid in Allylboration of Aldehydes: Catalytic, Regiospecific, Diastereospecific, and Enantioselective Synthesis of Homoallyl Alcohols. *J. Am. Chem. Soc.* 2002, *124*, 12414–12415.
9. Wada, R.; Oisaki, K.; Kanai, M.; Shibasaki, M. Catalytic Enantioselective Allylboration of Ketones. *J. Am. Chem. Soc.* 2004, *126*, 8910–8911.
10. Carosi, L.; Lachance, H.; Hall, D. G. Additions of Functionalized A-Substituted Allylboronates to Aldehydes under the Novel Lewis and Brønsted Acid Catalyzed Manifolds. *Tetrahedron* 2005, *46*, 8981–8985.
11. Yu, S. H.; Ferguson, M. J.; McDonald, R.; Hall, D. G. Brønsted Acid-Catalyzed Allylboration: Short and Stereodivergent Synthesis of All Four Eupomatilone Diastereomers with Crystallographic Assignments. *J. Am. Chem. Soc.* 2005, *127*, 12808–12809.
12. Rauniyar, V.; Hall, D. G. Catalytic Enantioselective and Catalyst-Controlled Diastereofacial-Selective Additions of Allyl- and Crotylboronates to Aldehydes Using Chiral Brønsted Acids. *Angew. Chem. Int. Ed.* 2006, *45*, 2426–2428.
13. Elford, T. G.; Arimura, Y.; Yu, S. H.; Hall, D. G. Triflic Acid-Catalyzed Additions of 2-Alkoxycarbonyl Allylboronates to Aldehydes. Study of Scope and Mechanistic Investigation of the Reaction Stereochemistry. *J. Org. Chem.* 2007, *72*, 1276–1284.

14. Rauniyar, V.; Zhai, H.; Hall, D. G. Catalytic Enantioselective Allyl- and Crotylboration of Aldehydes Using Chiral Diol•SnCl₄ Complexes. Optimization, Substrate Scope and Mechanistic Investigations. *J. Am. Chem. Soc.* 2008, *130*, 8481–8490.

15. Jain, P.; Antilla, J. C. Chiral Brønsted Acid-Catalyzed Allylboration of Aldehydes. *J. Am. Chem. Soc.* 2010, *132*, 11884–11886.

16. Jain, P.; Wang, H.; Houk, K. N.; Antilla, J. C. Brønsted Acid Catalyzed Asymmetric Propargylation of Aldehydes. *Angew. Chem. Int. Ed.* 2012, *51*, 1391–1394.

17. Grayson, M. N.; Pellegrinet, S. C.; Goodman, J. M. Mechanistic Insights into the BINOL-Derived Phosphoric Acid-Catalyzed Asymmetric Allylboration of Aldehydes. *J. Am. Chem. Soc.* 2012, *134*, 2716–2722.

18. Wang, H.; Jain, P.; Antilla, J. C.; Houk, K. N. Origins of Stereoselectivities in Chiral Phosphoric Acid Catalyzed Allylborations and Propargylations of Aldehydes. *J. Org. Chem.* 2013, *78*, 1208–1215.

19. Yamanaka, M.; Hoshino, M.; Katoh, T.; Mori, K.; Akiyama, T. Kinetic Resolution in Chiral Phosphoric Acid Catalyzed Aldol Reactions: Enantioselective Robinson-Type Annulation Reactions. *Eur. J. Org. Chem.* 2012, *24*, 4508–4514.

20. Itoh, J.; Fuchibe, K.; Akiyama, T. Chiral Phosphoric Acid Catalyzed Enantioselective Friedel–Crafts Alkylation of Indoles with Nitroalkenes: Cooperative Effect of 3 Å Molecular Sieves. *Angew. Chem. Int. Ed.* 2008, *47*, 4016–4018.

21. Hirata, T.; Yamanaka, M. DFT Study of Chiral-Phosphoric-Acid-Catalyzed Enantioselective Friedel-Crafts Reaction of Indole with Nitroalkene: Bifunctionality and Substituent Effect of Phosphoric Acid. *Chem. Asian J.* 2011, *6*, 510–516.

22. Jiang, G. J.; Wang, Y; Yu, Z. X. DFT Study on the Mechanism and Stereochemistry of the Petasis-Ferrier Rearrangements. *J. Org. Chem.* 2013, *78*, 6947–6955.

23. Kanomata, K.; Toda, Y.; Shibata, Y.; Yamanaka, M.; Tsuzuki, S.; Gridnev, I. D.; Terada, M. Secondary Stereocontrolling Interactions in Chiral Brønsted Acid Catalysis: Study of a Petasis-Ferrier-Type Rearrangement Catalyzed by Chiral Phosphoric Acids. *Chem. Sci.* 2014, *5*, 3515–3523.

24. Terada, M.; Soga, K.; Momiyama, N. Enantioselective Activation of Aldehydes by Chiral Phosphoric Acid Catalysts in an Aza-Ene-Type Reaction between Glyoxate and Enecarbamate. *Angew. Chem. Int. Ed.* 2008, *47*, 4122–4175.

25. Terada, M.; Tanaka, H.; Sorimachi, K. Enantioselective Direct Aldol-Type Reaction of Azlactone via Protonation of Vinyl Ethers by a Chiral Brønsted Acid Catalyst. *J. Am. Chem. Soc.* 2009, *131*, 3430–3431.

26. Momiyama, N.; Tabuse, H; Terada, M. Chiral Phosphoric Acid-Governed Anti-Diastereoselective and Enantioselective Hetero-Diels-Alder Reaction of Glyoxylate. *J. Am. Chem. Soc.* 2009, *131*, 12882–12883.

27. Maity, P.; Pemberton, R. P.; Tantillo, D. J.; Tambar, U. K. Brønsted Acid Catalyzed Enantioselective Indole Aza-Claisen Rearrangement Mediated by an Arene CH–O Interaction. *J. Am. Chem. Soc.* 2013, *135*, 16380–16383.

28. Monaco, M. R.; Prévost, S.; List, B. Catalytic Asymmetric Synthesis of Thiols. *J. Am. Chem. Soc.* 2014, *136*, 16982–16985.

29. Ajitha, M.; Huang, K.-W. Role of Keto–Enol Tautomerization in a Chiral Phosphoric Acid Catalyzed Asymmetric Thiocarboxylysis of Meso-epoxide: A DFT Study. *Org. Biomol. Chem.* 2015, *13*, 10981–10985.

30. Liao, S.; Ori, I.; Wang, Q.; List, B. Activation of H_2O_2 by Chiral Confined Brønsted Acids: A Highly Enantioselective Catalytic Sulfoxidation. *J. Am. Chem. Soc.* 2012, *134*, 10765–10768.

31. Jindal, G.; Sunoj, R. B. Axially Chiral Imidodiphosphoric Acid Catalyst for Asymmetric Sulfoxidation Reaction: Insights on Asymmetric Induction. *Angew. Chem. Int. Ed.* 2014, *53*, 4432–4436.

32. Hiemstra, H.; Wynberg, H. Addition of Aromatic Thiols to Conjugated Cycloalkenones, Catalyzed by Chiral β-Hydroxy Amines. A Mechanistic Study on Homogeneous Catalytic Asymmetric Synthesis. *J. Am. Chem. Soc.* 1981, *103*, 417–430.

33. Li, H.; Wang, Y.; Tang, L.; Deng, L. Highly Enantioselective Addition of Malonate and β-Ketoester to Nitroalkenes: Asymmetric C–C Bond Formation with New Bifunctional Organic Catalysts Based on Chinchona Alkaloids. *J. Am. Chem. Soc.* 2004, *126*, 9906–9907.

34. Wu, Y.; Singh, P. R.; Deng, L. Asymmetric Olefin Isomerization of Butenolides via Proton Transfer Catalysis by an Organic Molecule. *J. Am. Chem. Soc.* 2011, *133*, 12458–12461.

35. Xue, X. S.; Li, X.; Yu, A.; Yang, C.; Song, C.; Cheng, J. P. Mechanism and Selectivity of Bioinspired Chinchona Alcaloid Derivatives Catalyzed Asymmetric Olefin Isomerization: A Computational Study. *J. Am. Chem. Soc.* 2013, *135*, 7462–7473.

36. Chen, W.; Du, W.; Yue, L.; Li, R.; Wu, Y.; Ding, L.-S.; Chen, Y.-C. Organocatalytic Enantioselective Indole Alkylations of α,β-Unsaturated Ketones. *Org. Biomol. Chem.* 2007, *5*, 816–821.

37. Bartoli, G.; Carlone, A.; Pesciaioli, F.; Sambri, L.; Melchiorre, P. Organocatalytic Asymmetric Friedel–Crafts Alkylation of Indoles with Simple α,β-Unsaturated Ketones. *Org. Lett.* 2007, *9*, 1403–1405.

38. Nielsen, M.; Worgull, D.; Zweifel, T.; Gschwend, B.; Bertelsen, S.; Jørgensen, K. A. Mechanisms in Aminocatalysis. *Chem. Commun.* 2011, *47*, 632–649.

39. Moran, A.; Hamilton, A.; Bo, C.; Melchiorre, P. A Mechanistic Rationale for the 9-Amino(9-Deoxy)*Epi* Cinchona Alkaloids Catalyzed Asymmetric Reactions via Iminium Ion activation of Enones. *J. Am. Chem. Soc.* 2013, *135*, 9091–9098.

40. Jiang, H.; Sun, Y.; Liu, H.; Huang, X. Theoretical Study on Mechanism of Cinchona Alkaloids Catalyzed Asymmetric Conjugate Addition of Dimethyl Malonate to β-Nitrostyrene. *Int. J. Quant. Chem.* 2014, *114*, 642–651.

41. Kwiatkowski, P.; Beeson, T. D.; Conrad, J. C.; MacMillan, D. W. C. Enantioselective Organocatalytic α-Fluorination of Cyclic Ketones. *J. Am. Chem. Soc.* 2011, *133*, 1738–1741.

42. Lam, Y.-H.; Houk, K. N. How Cinchona Alkaloid-Derived Primary Amines Control Asymmetric Electrophilic Fluorination of Cyclic Ketones. *J. Am. Chem. Soc.* 2014, *136*, 9556–9559.

43. Baudequin, C.; Loubassou, J.-F.; Plaquevent, J.-C.; Cahard, D. Enantioselective Electrophilic Fluorination: A Study of the Fluorine-Transfer from Achiral N–F Reagents to Cinchona Alkaloids. *J. Fluorine Chem.* 2003, *122*, 189–193.

44. Zhou, J.; Wakchaure, V.; Kraft, P.; List, B. Primary-Amine-Catalyzed Enantioselective Intramolecular Aldolizations. *Angew. Chem. Int. Ed.* 2008, *47*, 7656–7658.

45. Lam, Y.-H.; Houk, K. N. Origins of Stereoselectivity in Intramolecular Aldol Reactions Catalyzed by Cinchona Amines. *J. Am. Chem. Soc.* 2015, *137*, 2116–2127.

46. Dolling, U. H.; Davis, P.; Grabowski, E. J. J. Efficient Catalytic Asymmetric Alkylations. 1. Enantioselective Synthesis of (+)-Indacrinone Via Chiral Phase-Transfer Catalysis. *J. Am. Chem. Soc.* 1984, *106*, 446–447.

47. Hughes, D. L.; Dolling, U. H.; Ryan, K. M.; Schoenewaldt, E. F.; Grabowski, E. J. J. Efficient Catalytic Asymmetric Alkylations. 3. A Kinetic and Mechanistic Study of the Enantioselective Phase-Transfer Methylation of 6,7-Dichloro-5-Methoxy-2-Phenyl-1-Indanone. *J. Org. Chem.* 1987, *52*, 4745–4752.

48. Corey, E. J.; Xu, F.; Noe, M. C. A Rational Approach to Catalytic Enantioselective Enolate Alkylation Using a Structurally Rigidified and Defined Chiral Quaternary Ammonium Salt under Phase Transfer Conditions. *J. Am. Chem. Soc.* 1997, *119*, 12414–12415.

49. Martins, E. F.; Pliego, R. Jr. Unraveling the Mechanism of the Cinchoninium Ion Asymmetric Phase-Transfer-Catalyzed Alkylation Reaction. *ACS Catalysis* 2013, *3*, 613–616.

50. Sigman, M. S.; Jacobsen, E. N. Schiff Base Catalysts for the Asymmetric Strecker Reaction Identified and Optimized from Parallel Synthetic Libraries. *J. Am. Chem. Soc.* 1998, *120*, 4901–4902.

51. Vachal, P.; Jacobsen, E. N. Structure-based Analysis and Optimization of a Highly Enantioselective Catalyst for the Strecker Reaction. *J. Am. Chem. Soc.* 2002, *124*, 10012–10014.

52. Schreiner, P. R.; Wittkopp, A. H-Bonding Additives Act Like Lewis Acid Catalysts. *Org. Lett.* 2002, *4*, 217–220.

53. Wittkopp, A.; Schreiner, P. R. Metal-Free, Noncovalent Catalysis of Diels–Alder Reactions by *Neutral* Hydrogen Bond Donors in Organic Solvents and in Water. *Chem. Eur. J.* 2003, *9*, 407–414.

54. Schreiner, P. R. Metal-free Organocatalysis through Explicit Hydrogen Bonding Interactions. *Chem. Soc. Rev.* 2003, *32*, 289–296.

55. Okino, T.; Hoashi, Y.; Takemoto, Y. Enantioselective Michael Reaction of Malonates to Nitroolefins Catalyzed by Bifunctional Organocatalysts. *J. Am. Chem. Soc.* 2003, *125*, 12672–12673.

56. Sohtome, Y.; Hashimoto, Y.; Nagasawa, K. Guanidine-Thiourea Bifunctional Organocatalyst for the Asymmetric Henry (Nitroaldol) Reaction. *Adv. Synth. Catal.* 2005, *347*, 1643–1648.

57. Shi, Y.-L.; Shi, M. Chiral Thiourea-Phosphine Organocatalysts in the Asymmetric Aza-Morita–Baylis–Hillman Reaction. *Adv. Synth. Catal.* 2007, *349*, 2129–2135.

58. Okino, T.; Hoashi, Y.; Furukawa, T.; Xu, X.; Takemoto, Y. Enantio-and Diastereoselective Michael Reaction of 1,3-Dicarbonyl Compounds to Nitroolefins Catalyzed by a Bifunctional Thiourea. *J. Am. Chem. Soc.* 2005, *127*, 119–125.

59. Hammar, P.; Marcelli, T.; Hiemstra, H.; Himo, F. Density Functional Theory Study of the Cinchona Thiourea-Catalyzed Henry Reaction: Mechanism and Enantioselectivity. *Adv. Synth. Catal.* 2007, *349*, 2537–2548.

60. Zuend, S. J.; Jacobsen, E. N. Cooperative Catalysis by Tertiary Amino-Thioureas: Mechanism and Basis for Enantioselectivity of Ketone Cyanosilylation. *J. Am. Chem. Soc.* 2007, *129*, 15872–15883.

61. Zhang, Z. G.; Schreiner, P. R. (Thio)urea Organocatalysis—What Can Be Learnt from Anion Recognition? *Chem. Soc. Rev.* 2009, *38*, 1187–1198.

62. Hamza, A.; Schubert, G.; Soós, T.; Pápai, I. Theoretical Studies on the Bifunctionality of Chiral Thiourea-Based Organocatalysts: Competing Routes to C-C Bond Formation. *J. Am. Chem. Soc.* 2006, *128*, 13151–13160.

63. Zuend, S. J.; Jacobsen, E. N. Mechanism of Amido-Thiourea Catalyzed Enantioselective Imine Hydrocyanation: Transition State Stabilization via Multiple Non-Covalent Interactions. *J. Am. Chem. Soc.* 2009, *131*, 15358–15374.

64. Zhang, Y.; Shao, Y.-L.; Xu, H.-S.; Wang, W. J. Organocatalytic Direct Asymmetric Vinylogous Michael Reaction of an α,β-Unsaturated γ-Butyrolactam with Enones. *Org. Chem.* 2011, *76*, 1472–1474.

65. Zhu, J.-L.; Zhang, Y.; Liu, C.; Zheng, A.-M.; Wang, W. Insights into the Dual Activation Mechanism Involving Bifunctional Cinchona Alkaloid Thiourea Organocatalysts: An NMR and DFT Study. *J. Org. Chem.* 2012, *77*, 9813–9825.

66. Amere, M.; Lasne, M. C.; Rouden, J. Highly Enantioselective Decarboxylative Protonation of α-Aminomalonates Mediated by Thiourea Cinchona Alkaloid Derivatives: Access to Both Enantiomers of Cyclic and Acyclic α-Aminoacids. *Org. Lett.* 2007, *9*, 2621.

67. Blanchet, J.; Baudoux, J.; Amere, M.; Lasne, M. C.; Rouden, J. Asymmetric Malonic and Acetoacetic Acid Syntheses—A Century of Enantioselective Decarboxylative Protonations. *Eur. J. Org. Chem.* 2008, 5493.

68. Sengupta, A.; Sunoj, R. B. Mechanistic Insights on Organocatalytic Enantioselective Decarboxylative Protonation by Epicinchona-Thiourea Hybrid Derivatives. *J. Org. Chem.* 2012, *77*, 10525–10536.

69. Kitagaki, S.; Ueda, T.; Mukai, C. Planar Chiral [2.2]Paracyclophane-Based Bis(thiourea) Catalyst: Application to Asymmetric Henry Reaction. *Chem. Commun.* 2013, *49*, 4030–4032.

70. Breugst, M.; Houk, K. N. Computational Analysis of Cyclophane-Based Bisthiourea-Catalyzed Henry Reactions. *J. Org. Chem.* 2014, *79*, 6302–6309.

71. Odagi, M.; Furukori, K.; Watanabe, T.; Nagasawa, K. Asymmetric α-Hydroxylation of Tetralone-Derived β-Ketoesters by Using a Guanidine–Urea Bifunctional Organocatalyst in the Presence of Cumene Hydroperoxide. *Chem. Eur. J.* 2013, *19*, 16740–16745.

72. Odagi, M.; Furukori, K.; Yamamoto, Y.; Sato, M.; Iida, K.; Yamanaka, M.; Nagasawa, K. Origin of Stereocontrol in Guanidine-Bisurea Bifunctional Organocatalyst That Promotes α-Hydroxylation of Tetralone-Derived β-Ketoesters: Asymmetric Synthesis of β- and γ-Substituted Tetralone Derivatives via Organocatalytic Oxidative Kinetic Resolution. *J. Am. Chem. Soc.* 2015, *137*, 1909–1915.

73. Corey, E. J.; Bakshi, R. K.; Shibata, S. Highly Enantioselective Borane Reduction of Ketones Catalyzed by Chiral Oxazaborolidines. Mechanism and Synthetic Implications. *J. Am. Chem. Soc.* 1987, *109*, 5551–5553.

74. Corey, E. J.; Bakshi, R. K.; Shibata, S.; Chen, C. P.; Singh, V. K. A Stable and Easily Prepared Catalyst for the Enantioselective Reduction of Ketones. Applications to Multistep Syntheses. *J. Am. Chem. Soc.* 1987, *109*, 7925–7926.

75. Demir, A. S.; Sesenoglu, O.; Aksoy-Cam, H.; Kaya, H.; Aydogan, K. Enantioselective Synthesis of Both Enantiomers of 2-Amino-2-(2-Furyl) Ethan-1-ol as a Flexible Building Block for the Preparation of Serine and Azasugars. *Tetrahedron Asymmetry* 2003, *14*, 1335–1340.

76. Gosselin, F.; O'Shea, P. D.; Roy, S.; Reamer, R. A.; Chen, C. Y.; Volante, R. P. Unprecedented Catalytic Asymmetric Reduction of N-H Imines. *Org. Lett.* 2005, *7*, 355–358.

77. Ryu, D. H.; Zhou, G.; Corey, E. J. Enantioselective and Structure-Selective Diels-Alder Reactions of Unsymmetrical Quinones Catalyzed by a Chiral Oxazaborolidinium Cation. Predictive Selection Rules. *J. Am. Chem. Soc.* 2004, *126*, 4800–4802.

78. Shibatomi, K.; Futatsugi, K.; Kobayashi, F.; Iwasa, S.; Yamamoto, H. Stereoselective Construction of Halogenated Quaternary Stereogenic Centers via Catalytic Asymmetric Diels-Alder Reaction. *J. Am. Chem. Soc.* 2010, *132*, 5625–5627.

79. Canales, E.; Corey, E. J. Highly Enantioselective [2 + 2]-Cycloaddition Reactions Catalyzed by a Chiral Aluminum Bromide Complex. *J. Am. Chem. Soc.* 2007, *129*, 12686–12687.

80. Liu, D.; Hong, S. W.; Corey, E. J. Enantioselective Synthesis of Bridged- or Fused-Ring Bicyclic Ketones by a Catalytic Asymmetric Michael Addition Pathway. *J. Am. Chem. Soc.* 2006, *128*, 8160–8161.

81. Wang, Y.; Guo, X.; Tang, M.; Wei, D. Theoretical Investigations toward the Asymmetric Insertion Reaction of Diazoester with Aldehyde Catalyzed by N-Protonated Chiral Oxazaborolidine: Mechanisms and Stereoselectivity. *J. Phys. Chem. A* 2015, *119*, 8422–8431.

82. Gao, L.; Kang, B. C.; Ryu, D. H. Catalytic Asymmetric Insertion of Diazoesters into Aryl-CHO Bonds: Highly Enantioselective Construction of Chiral All-Carbon Quaternary Centers. *J. Am. Chem. Soc.* 2013, *135*, 14556–14559.

83. Stephan, D. W. Frustrated Lewis Pairs: From Concept to Catalysis. *Acc. Chem. Res.* 2015, *48*, 306–316.

84. Stephan, D. W.; Erker, G. Frustrated Lewis Pair Chemistry of Carbon, Nitrogen and Sulfur Oxides. *Chem. Sci.* 2014, *5*, 2625–2641.

85. Stephan, D. W.; Greenberg, S.; Graham, T. W.; Chase, P.; Hastie, J. J.; Geier, S. J.; Farrell, J. M.; Brown, C. C.; Heiden, Z. M.; Welch, G. C.; Ullrich, M. Metal-Free Catalytic Hydrogenation of Polar Substrates by Frustrated Lewis Pairs. *Inorg. Chem.* 2011, *50*, 12338–12348.

86. Stephan, D. W.; Erker, G. Frustrated Lewis Pairs: Metal-Free Hydrogen Activation and More. *Angew. Chem. Int. Ed.* 2010, *49*, 46–76.

87. Lindqvist, M.; Borre, K.; Axenov, K.; Kótai, B.; Nieger, M.; Leskelä, M.; Pápai, I.; Repo, T. Chiral Molecular Tweezers: Synthesis and Reactivity in Asymmetric Hydrogenation. *J. Am. Chem. Soc.* 2015, *137*, 4038–4041.

88. Zhang, Z.; Du, H. A Highly cis-Selective and Enantioselective Metal-Free Hydrogenation of 2,3-Disubstituted Quinoxalines. *Angew. Chem. Int. Ed.* 2015, *54*, 623–626.

89. Wei, S.; Du, H. A Highly Enantioselective Hydrogenation of Silyl Enol Ethers Catalyzed by Chiral Frustrated Lewis Pairs. *J. Am. Chem. Soc.* 2014, *136*, 12261–12264.

90. Liu, Y.; Du, H. Chiral Dienes as "Ligands" for Borane-Catalyzed Metal-Free Asymmetric Hydrogenation of Imines. *J. Am. Chem. Soc.* 2013, *135*, 6810–6813.

91. Ghattas, G.; Chen, D.; Pan, F.; Klankermayer, J. Asymmetric Hydrogenation of Imines with a Recyclable Chiral Frustrated Lewis Pair Catalyst. *Dalton Trans.* 2012, *41*, 9026–9028.
92. Sumerin, V.; Chernichenko, K.; Nieger, M.; Leskelä, M.; Rieger, B.; Repo, T. Highly Active Metal-Free Catalysts for Hydrogenation of Unsaturated Nitrogen-Containing Compounds. *Adv. Synth. Catal.* 2011, *353*, 2093–2110.
93. Chen, D.; Wang, Y.; Klankermayer, J. Enantioselective Hydrogenation with Chiral Frustrated Lewis Pairs. *Angew. Chem. Int. Ed.* 2010, *49*, 9475–9478.
94. Battaglia, M. R.; Buckingham, A. D.; Williams, J. H. The Electric Quadrupole Moments of Benzene and Hexafluorobenzene. *Chem. Phys. Lett.* 1981, *78*, 421–423.

Conclusions

The main message that the authors would like to share with the readers of this book is as follows: *Elucidation of the intrinsic mechanism of generation of chirality in catalytic asymmetric reactions is a much more ambiguous task than it is commonly accepted.* That is stipulated for two main reasons: (1) experimental problems in the studies of catalytic cycles and (2) inevitable ambiguity in the interpretation of the computational data.

The reaction pool of a catalytic asymmetric reaction is a sophisticated system involving numerous species that are either involved in a catalytic cycle or are off-loop species. All these species are likely to coexist in fast equilibria, which opens a possibility for various pathways to converge or return to the resting state before the catalytic cycle is completed, yielding the product and recovering the catalyst. Solvent molecules are likely to participate in these equilibria. A vast majority of these interconverting species cannot be detected experimentally due to their relative instability and/or high reactivity.

Commonly applied models of reaction kinetics, such as the transition state theory or RRKM (Rice–Ramsperger–Kassel–Marcus) theory, cannot completely describe multistage catalytic cycles or even the simplest reactions in many cases. As a catalytic cycle is a nonlinear dynamic system, statements such as "the reaction is first order with respect to a component A" or "the reaction is zero order with respect to a component B" cannot be universally assumed.

In the case of intermolecular stages, it is important to consider the importance of the process that brings the reagents together before the reactive adduct is actually formed in a proper conformation. Several examples are described in this book (e.g., Rh-catalyzed asymmetric hydrogenation of enamides or Ir-catalyzed asymmetric hydrogenation of exocyclic α,β-unsaturated carbonyl compounds) that suggest that extremely high levels of enantioselection are observed in the cases when the formation of a reactive adduct is efficiently blocked for one of the diastereomeric catalytic pathways.

Furthermore, most of the elementary steps that occur in the reaction pool of a catalytic asymmetric reaction are stereoselective, because usually intermediates are diastereomers having different chemical properties. It means that if we are able to observe a stereoselective transformation of some species experimentally, we cannot directly conclude that this transformation is a stereodetermining stage of the catalytic cycle, even if this transformation can lead to the correct handedness of the product.

In the same way, if we are able to compute some stage of a catalytic cycle to be stereoselective and providing the expected handedness, without detailed studies of the whole catalytic cycle, we cannot be entirely sure that this stage determines its stereochemical outcome.

Moreover, it has been recognized recently that nonconventional weak intramolecular interactions must play an important role in the stabilization of intermediates and transition states, thus strongly affecting the process of enantioselection. However, important features of these interactions, that is, relative strength, effective range, directionality, etc., are practically unknown. It makes impossible at the moment real accounting of numerous intramolecular effects present in any intermediate or transition state, that in turn discredits the claims for "conscious catalyst design."

For example, it is known that increasing the size of the catalyst often results in higher optical yields. However, it is impossible at the moment to attribute this effect to any particular kind of intramolecular interactions, because along with the evident increase of "steric bulkiness," the network of possible nonconventional attractive interactions is also increasing.

All complications listed earlier lead to the conclusion that computational studies of asymmetric catalytic reactions with low enantiometric excess (ee's) (less than 90% ee at the very least) are not expected to provide any data that could be reasonably rationalized. A low optical yield may just mean that several pathways are competing, and not necessarily those differing only in the handedness of the product.

In contrast, a high optical yield (especially 99% ee or higher) allows us to conclude reliably that all conceivable pathways irrespective of their numbers are "cut off," and there is only one reaction pathway that leads to the product. This certitude can bring plenty of new chemical knowledge upon careful analysis of the mechanism of exclusion of all other possible pathways from the flux of catalysis.

Such analysis should avoid any simplifications that inevitably appear when the researches attempt to describe the process of enantioselection in terms of empirical qualitative schemes. All conceivable catalytic pathways should be computed at the highest possible level of theory for the real molecules and accounting for the solvent effects and possible convergence of the pathways. After explicit simulation of kinetics using the computed thermodynamic parameters, reliable data on the influence of the structural parameters of the catalyst on the sophisticated process of enantioselection can be potentially obtained. From these data, one can hope to obtain the basis for calibration of various types of intramolecular interactions with respect to their strength and operating distances.

In other words, we are now not in a position to "rationalize the catalyst design," because we do not have sufficiently reliable data to do this. Hence, mechanistic studies of the chiral catalytic reactions should mainly pursue an opposite task: accumulate and calibrate data on the weak intramolecular interactions. This should preferably be done with reactions in which the ee is higher than 99%. If we could understand what features of a catalytic cycle are essential to give the product with 99% ee, we could begin to consider how "to make a better catalyst."

In conclusion, one must admit that so far the effective chiral catalysts are being developed through hard work of the students working on a trial-and-error basis rather than by clever mechanistic ideas or computational design of their professors.

Index

Note: Page numbers followed by f, t and c refer to figures, tables and chart respectively. Page number in italics refer to schemes.

Printed and bound by CPI Group (UK) Ltd, Croydon, CR0 4YY
01/11/2024
01782617-0003